Palgrave Studies in Literature, Science and Medicine

Series Editors:

Sharon Ruston
Lancaster University
Lancaster, United Kingdom

Alice Jenkins
University of Glasgow
Glasgow, United Kingdom

Catherine Belling
Northwestern University
Chicago, USA

Palgrave Studies in Literature, Science and Medicine is an exciting new series that focuses on one of the most vibrant and interdisciplinary areas in literary studies: the intersection of literature, science and medicine. Comprised of academic monographs, essay collections, and Palgrave Pivot books, the series will emphasize a historical approach to its subjects, in conjunction with a range of other theoretical approaches. The series will cover all aspects of this rich and varied field and is open to new and emerging topics as well as established ones. Editorial board: Steven Connor, Professor of English, University of Cambridge, UK Lisa Diedrich, Associate Professor in Women's and Gender Studies, Stony Brook University, USA Kate Hayles, Professor of English, Duke University, USA Peter Middleton, Professor of English, University of Southampton, UK Sally Shuttleworth, Professorial Fellow in English, St Anne's College, University of Oxford, UK Susan Squier, Professor of Women's Studies and English, Pennsylvania State University, USA Martin Willis, Professor of English, University of Westminster, UK

More information about this series at
http://www.springer.com/series/14613

Martin Willis
Editor

Staging Science

Scientific Performance on Street, Stage and Screen

Editor
Martin Willis
Department of English, Communication & Philosophy
Cardiff University, Cardiff
United Kingdom

Palgrave Studies in Literature, Science and Medicine
ISBN 978-1-137-49993-6   ISBN 978-1-137-49994-3   (eBook)
DOI 10.1057/978-1-137-49994-3

Library of Congress Control Number: 2016942402

Printed on acid-free paper

This Palgrave Pivot imprint is published by Springer Nature
The registered company is Macmillan Publishers Ltd. London

ACKNOWLEDGEMENTS

This collection of essays would not have had its impetus without the support of the University of Westminster's Department of English, which provided the funding for a colloquium on scientific performance held in London in December 2013. The colloquium and its surrounding events—particularly the re-creation of the Pepper's Ghost illusion which formed a centre-piece for the activities—would have been impossible to mount without the initial inspiration and continuing enthusiasm of the Head of the Department of English, Alexandra Warwick. This collection is dedicated to her, with gratitude.

CONTENTS

1 Introduction: Imaginative Mobilities   1
  Martin Willis

2 Making the Most Beautiful Experiment: Reconstructing
  Gassiot's Cascade   11
  Iwan Rhys Morus

3 Science in the City: Scientific Display and Urban
  Performance in Victorian Travel Guides to London   35
  Martin Willis

4 Of Hats and Scientific Laughter   59
  Tiffany Watt Smith

5 'You Can't Make a Film About Mice Just by
  Going Out into a Meadow and Looking at Mice':
  Staging as Knowledge Production in Natural History
  Film-making   83
  Jean-Baptiste Gouyon

6 'Unmediated' Science Plays: Seeing What Sticks 105
Kirsten E. Shepherd-Barr

Afterword 125
Bernard Lightman

Index 133

NOTES ON CONTRIBUTORS

Jean-Baptiste Gouyon is a historian of science with interests in the presentation and uses of science in visual media. His research is on science in films and on television, with a focus on wildlife documentaries. Currently, he is a teaching fellow at University College London Department of Science and Technology Studies.

Bernard Lightman is Professor of Humanities at York University, and Vice-President (2016–2017) and President (2018–2019) elect of the History of Science Society. Lightman's research interests include nineteenth-century popular science and Victorian scientific naturalism. Among his most recent publications are *Victorian Popularizers of Science, Victorian Scientific Naturalism* (co-edited with Gowan Dawson), *Evolution and Victorian Culture* (co-edited with Bennett Zon), and *The Age of Scientific Naturalism* (co-edited with Michael Reidy). He is currently working on a biography of John Tyndall and is one of the editors of the John Tyndall Correspondence Project, an international collaborative effort to obtain, digitalize, transcribe, and publish all surviving letters to and from Tyndall.

Iwan Rhys Morus is Professor of History at the University of Aberystwyth and has published extensively on the history and culture of Victorian science. His most recent book is *Shocking Bodies: Life, Death & Electricity in Victorian England* (2011). He is currently writing a biography of the Welsh natural philosopher William Robert Grove and a cultural history of Victorian illusions, as well as editing the *Oxford Illustrated History of Science*. His current research looks at the cultural history of vision in the nineteenth century and the relationship between the culture of scientific spectacle and the constitution of epistemic authority. He is presently co-investigator on the AHRC funded project 'Unsettling Scientific Stories: Expertise, Narrative and Future Histories', and is also the Project Director for the 'Memory and Media in Wales' JISC-funded research project as well as a senior collaborator on the John Tyndall Correspondence Project at Montana State

University. He is on the Editorial Board of the *British Journal of the History of Science* and was the editor of *History of Science* until the end of 2014. He is currently on the Editorial Board of the University of Wales Press series, Scientists of Wales.

Kirsten E. Shepherd-Barr is Professor of English and Theatre Studies at the University of Oxford and a Fellow of St Catherine's College. She is the author of *Theatre and Evolution from Ibsen to Beckett* (2015), *Science on Stage: From Doctor Faustus to Copenhagen* (2006; paperback 2012), and *Ibsen and Early Modernist Theatre, 1890–1900* (1997), as well as *Modern Drama: A Very Short Introduction* (2016). She co-edited two special issues of the journal *Interdisciplinary Science Reviews* on 'New Directions in Theatre and Science' (2013 and 2014).

Tiffany Watt Smith is a Lecturer in Drama, Theatre and Performance at Queen Mary University of London, and currently holds a Wellcome Trust funded research fellowship at the Centre for the History of the Emotions. Before beginning her career in research she worked as a professional theatre director for 10 years, including as International Associate Director at the Royal Court, and won the Jerwood Award for Directors at the Young Vic in 2004. Her publications include *On Flinching: Theatricality and Scientific Looking from Darwin to Shell Shock* (2014), and *The Book of Human Emotions* (UK; 2015, US; 2016).

Martin Willis is Professor of English Literature at Cardiff University, Chair of the British Society for Literature and Science (2015–2018) and editor of the *Journal of Literature and Science*. His previous works include *Literature and Science* (Palgrave, 2015) and the award-winning *Vision, Science and Literature, 1870–1920: Ocular Horizons* (2011). He has also written on various topics related to Victorian literature, science and medicine and is presently working towards a book on science and the Victorian travel guide as well as leading a project on seizure-inducing diseases from the early nineteenth century to the present.

LIST OF FIGURES

Fig. 4.1   Hercat, *Chapeaugraphy, Shadowgraphy, and Paper-Folding*
             (London: Fleet Street, 1909). In the collection of the author.   69
Fig. 4.2   Hercat, *Chapeaugraphy, Shadowgraphy, and Paper-Folding*
             (London: Fleet Street, 1909). In the collection of the author.   70

CHAPTER 1

Introduction: Imaginative Mobilities

Martin Willis

Abstract Martin Willis argues that the work collected together in this collection offers new ways of conceiving of scientific performance. While science in action has been extensively studied, there remain other kinds of questions that have still to be answered. In particular the study of scientific performances from a range of perspectives will illuminate the relationships between the sciences and imagination, as well as between the aesthetic and the epistemic. Focusing on scientific performance across a range of humanities disciplines—literary studies, media studies, drama, and history—additionally reveals the strength of interdisciplinary methodologies within the humanities rather than between the humanities and the sciences. This collection may, then, be regarded as an effort to showcase a new science humanities.

This volume began life as a series of events—also titled Staging Science— held across 2 days at the University of Westminster in December 2013. Both the location and date are important, as the events were organised around a re-staging of Henry Pepper's ghost illusion first performed at the University of Westminster when it was the Royal Polytechnic

M. Willis
Cardiff University, UK

© The Editor(s) (if applicable) and The Author(s) 2016
M. Willis (ed.), *Staging Science*,
DOI 10.1057/978-1-137-49994-3_1

1

Institution in the 1860s. The re-staging was undertaken by Richard Hand and Geraint D'Arcy, both academics working in the Faculty of Creative Industries at the University of South Wales in Cardiff. It was, in fact, their Faculty Research Director, Ruth McElroy, who first suggested the collaboration and believed Richard and Geraint would be ideal contributors to Staging Science. So it proved. Their digital updating of the Pepper's ghost illusion, along with their adaptation of Charles Dickens's short story 'The Haunted Man' (which Pepper had used in his original script) came together to provide the audience with an uncanny return to mid-Victorian scientific performance. Readers interested in seeing the re-staging can still get a taste of it in an online video.[1] Pepper's ghost provided a potent central symbol for the other events—a round-table discussion of science and performance and a symposium—in which the contributors to this volume took part. The events, and especially the re-staging of Pepper's Ghost, spoke particularly to the centrality of performance in presenting science to new audiences. It is in response to the significance of this that this present volume brings together scholars from the disciplines of literature, history, and media studies to investigate the different methods and meanings of performing science. Most analyses of science as praxis ask what it is that science, and science scholarship, gains from the altered perspective that performativity provides. This is explicitly the case, for example, in the ground-breaking work of science studies scholars Bruno Latour and Andrew Pickering, who ask what it means to science to think through practice as well as what it might mean for science and technology studies to look at scientific action through this lens.[2] Similarly, in the influential work on popular science by Jim Secord and Jonathan Topham, it is both what science accrues by thinking of popular and elite science as part of a continuum and what history of science criticism might learn from rethinking outside of diffusionist understandings of scientific knowledge-making that is under discussion.[3] This collection of essays builds on that work by asking complementary but different questions. What kind of knowledge does performance elicit, and for whom? Who performs and for what reasons? In what ways does performance alter the relationship between science and its audiences? What relationships exist between the aesthetic and the epistemic, and between the imagination and the sciences? For all the writers here, such questions are at the heart of their investigations into the staging of science regardless of period or form. The significance of their work, while apparent in each chapter, also comes from the cross-fertilisation of the ideas and methods of multiple disciplines. Indeed, it is that multiplicity that the volume offers as a new approach to science and performance; it is one that connects the different disciplines of the humanities in a fresh form of critical encounter with science. Although not (yet) a commonly used phrase, the volume may well be described as a pathfinder for a 'science humanities' that endeavours to think carefully about science through the multivalent knowledges that the humanities can bring to bear.

In turn, the questions offered above emerge from a specific principle about the role of imaginative performance in the sciences: that it is the imagination that, at least in part, enables knowledge-production to occur and which is far too often obscured or erased from stories of scientific understanding. John Tyndall, for example, to whom Bernard Lightman returns in his Afterword to this collection, understood the importance of the imagination in scientific work, but ascribed to it, ultimately, a role only as a mode of transport towards a place where knowledge might be uncovered by rational investigation.[4] Similarly, in the twentieth century, Peter Medawar drew attention to the importance of the imagination in originating speculations about the world but, like Tyndall, saw it only as a vehicle taking the scientist towards more mature evidence-based theories.[5] To take issue with Tyndall and Medawar is a matter of emphasis rather than outright difference. Both, in complementary ways, acknowledge the imagination as a vital spark (I draw on Shelley's novel, *Frankenstein*, consciously here) in the early stages of scientific investigation.[6] Yet, in the end, they both also see the imagination as a handmaiden to science and its rational discoveries. It is the latter position which this collection interrupts, disrupts, and overthrows, and which a science humanities would continue to reject.

Indeed, central to the collection is the belief that the imagination—effected through and as performance—has far greater force in mobilising science than has previously been accepted and that this deserves further study. Mobility is the key word here and it requires some additional definition in this context. For the writers in this collection, mobility has many features: it registers the movement of scientific knowledge from one group to another; it tracks the exchange of ideas between performative and scientific sites and actors; and it notes how science itself is transformed into performance and vice versa. What is elided in thinking about such imaginative mobilities is a hierarchy of different realms of knowledge, most often resulting in the prioritising of scientific knowledge-making over the epistemes of performance. The inspiration for thinking through the idea of mobility can be found in Secord's important call for the renegotiation of the relationship between elite and popular science, where he argued that historians should be looking at 'knowledge in transit' rather than the top-down bestowing of knowledge from a privileged arena to one without power or status.

Other scholars have also made important interventions into the realm of performance, even when their focus has ostensibly been elsewhere. The work of historian of science Gerald Holton in the 1970s, for example, stressed the imaginative underpinning of three aspects of scientific work— the visual, analogical, and thematic—which, taken together, have clear connections to some of the ideals of imaginative performance: the necessity of creating something visually striking for an audience, of showing rather than telling, and of making a performance that is cohesive. In a very different way the work on seventeenth-century scientific writing and its dissemination by Steven Shapin and Simon Schaffer in *Leviathan and the Air Pump* (1985) revealed how works of science often used pictorial illustrations of new scientific instruments in action to promote their recognition by readers.[7] These illustrations were, in fact, imagined performances captured on the page and their importance to the new science illuminates the centrality of performance to the proper dissemination of new knowledge.

More recently and with a focus decidedly on issues of performance, Iwan Rhys Morus, whose work appears in this collection, gathered together a group of historians of science to reflect in the pages of the journal *Isis* on how performance might be studied anew. It is there that Morus suggests how the work of performance studies scholars might be employed to give alternative perspectives on the place of performance in science.[8] At the same time, Kirsten Shepherd-Barr, another writer to feature here, and Carina Bartleet conducted a new examination of science in the theatre across two issues of *Interdisciplinary Science Reviews*.[9] For them, this new work had emerged out of a desire to see how new performance modes had energised theatrical practice as well as how this had led to further cross-fertilisations between science and theatre. Equally influential has been the work of Sadiah Qureshi on scientific display, most forcefully in her 2011 book *Peoples on Parade*. Qureshi, taking a slightly different perspective, revealed how human exhibitions and performances were vital in the formation of sciences like anthropology.[10]

What, then, might performance studies offer to the historian of science, the science studies scholar, and the literature and science expert? As itself an interdiscipline, formed between drama and cultural studies, performance studies is sympathetically aligned to the ongoing projects of these scholarly groupings. Performance studies also has a distinguished history (reaching back to the work of critics such as Richard Schechner in the 1960s) in university theatre studies departments seeking new ways to articulate the different kinds of performances developing around them. In particular performance studies extended the range of what could be regarded as performance, looking not only at artistic performances conducted in authorised spaces such as theatres but beyond them at performances embedded in everyday life that took place on the streets, at work, or in the home. These cultural performances not only took place within cultural locations they also represented and even re-enacted cultural forms. Drawing on the sociological work on everyday performance led by Erving Goffman, performance studies scholars began to examine cultural performances much more extensively, and developed a vocabulary and a set of methodological principles for doing so.[11]

Performance became, across the 1980s and 1990s, an organising principle for studying a wide range of behaviours and situations as well as considering their aesthetic and epistemic content. Schechner saw this work as 'a means of understanding historical, social, and cultural processes'.[12] In opening out its interests to this extent, performance studies shares a set of values with historians of science and literature and science scholars who also wish to understand history, society, and culture. Increasingly, then, performance studies became another interdiscipline determined to investigate the creative historical processes that formed modern culture and society.

To do so, performance studies scholars focussed attention on three key aspects of cultural performances: event, agency, and embodiment. First, they asked questions of the meaning of individual events, including the significance of their location, timing, and explicit or implicit aims. Of particular interest were performances that might be difficult to repeat and which drew power from their lack of rehearsal. For Schechner the performative aspect of impromptu gatherings, for example, emerged from 'the possibility of improvisation' where 'the unexpected might happen'.[13] Second, performance studies asked important questions of the agency of the actors and, vitally, their audiences. Indeed a point was reached where all those present at cultural performances were regarded as actors. Erin Striff, for example, argues that in many cultural performances 'the performers themselves become the text to be read' while audience members often play important roles as inquisitors of performance and even co-creators of performative activities.[14] Third, the meanings associated with the body and with other material objects became central to understanding performances fully. The body was recognised as a site of performativity that can be analysed in combination with the other objects with which they interact. As Barbara Kirshenblatt-Gimblett argued, 'object performance provides a particularly rich arena for the relationship between people and things'.[15]

One further defining principle of much of the leading work in contemporary performance studies is a political commitment to celebrate and promote diversity, equality, and understanding. Performance studies, therefore, is clearly on what might be called the political left. Certainly it defines itself in opposition to global conservatism and to reactionary politics of any kind. In being so politically transparent, performance studies also has a great deal in common with the projects of historical scholarship and with (some) aspects of literature and science studies. There is surely interesting future research to be done which might draw on some of the strengths of performance studies to inculcate a greater political agency in the study of the relationship between imaginative and scientific activities.

Performance studies certainly offers the scholar of scientific performance significant tools for investigation. As Morus argues in his chapter here, thinking through non-textual means might be a profitable way to pursue future research on Victorian scientific demonstration. The relations that performance studies sees between people and things offers examples for this kind of work. Drawing insight from performance studies scholarship that considers performers and audiences as texts to be read is suggestive too for my own work on the representation of scientific sites in travel guides, which forms the second chapter of this collection. Meanwhile Watt Smith's chapter already acknowledges the centrality of performance within different social spheres in her work on laughter across comic performance and psychological experiments. Similarly, Shepherd-Barr's work on new types of theatrical experience which disrupt traditional notions of theatrical agency and narrative structure chimes with the foundations of performance studies. Finally, Gouyon's chapter on television documentary asks fascinating questions about the embodied nature of performance of both human and non-human actors.

These five very different articulations of the mobility of the imagination through scientific performance are a starting point for looking long and closely at the role of the imaginative performance in scientific activity: whether through the actions of scientific actors or in fiction and drama or in other media such as film and television. There is space to look further back than we do in this collection—to the performances of natural philosophy in the seventeenth century—but also to extend into the contemporary worlds of digital media and technology. There is, then, a much longer history of the relationships between science and the imagination to be written and a much greater attention to be paid to the diverse arenas where those collisions and collaborations occur. It seems timely for scholarship across several disciplines to approach the topic: for this project to be one way of beginning to think of a science humanities rather than in separate disciplinary locations. Never before has scientific performance combined the creative imagination with scientific knowledge in such diverse ways. The multi-channel world of contemporary television, for example, has seen the rise of specialist channels screening and now producing various kinds of science television. This has translated, at least in Britain, into mainstream terrestrial broadcasting of science on a very large scale indeed, and with extensive audience take-up. Recent years have also witnessed the rise of science stand-up and other forms of comedic performance with science as a foundation. On radio, television, and in live performance, science has a place in comedy that appears entirely new. Popular science writing is at a significant peak: creative non-fiction of this sort often reaches towards the top of the bestseller charts, and also has credibility as a distinct genre of writing. This is supported in no small part by significant annual competitions and prizes for the best popular science writing. In the UK such competitions are led by the Wellcome Trust, who also promote creative scientific performance through their ongoing series of artistic exhibitions in central London.[16] In North America, collaborations between science and art are even more numerous and long-standing; evidence can be found even in the architecture of scientific sites such as the Faculty of Science Centennial Centre for Interdisciplinary Science at the University of Alberta where the mosaic floor was produced as an art-science collaborative project.[17]

In popular media forms and via research institutions science draws upon the imagination to perform for diverse audiences. In what other ways might these aesthetic productions also produce knowledge, and what role does the imagination play in them doing so? In sum, how does imaginative praxis mobilise science? These are the key questions posed and answered by the writers here. They conduct their work, of course, in specific contexts. What the answers might be in other arenas is for future research. We hope that this collection might give impetus to scholars to take up the challenge, focus their attention on other combinations of science, performance, and imagination, and provide new ways of envisaging science's performativity.

NOTES

1. 'Pepper's Ghost Returns to 309 Regent Street', *YouTube*, accessed via http://www.youtube.co.uk
2. Bruno Latour, *Science in Action: How to Follow Scientists and Engineers through Society* (Cambridge: Harvard University Press, 1987); Andrew Pickering, *The Mangle of Practice: Time, Agency, and Science* (Chicago: University of Chicago Press, 1995).
3. James A. Secord, 'Knowledge in Transit', *Isis*, 95 (2004), 654–672; Jonathan Topham, 'Focus: Historicizing "Popular Science": Introduction', *Isis*, 100 (2009), 310–318.
4. John Tyndall, *On the Scientific Use of the Imagination: A Discourse* (London: Longman, Green & Co., 1870).
5. Peter Medawar, 'Science and Literature', *Encounter*, 32.1 (1969), 15–23.
6. Mary Shelley, *Frankenstein; Or, The Modern Prometheus* (1818; London: Penguin, 1992).
7. Steven Shapin and Simon Schaffer, *Leviathan and the Air-Pump: Hobbes, Boyle, and the Experimental Life* (Princeton: Princeton University Press, 1985).
8. Iwan Rhys Morus, 'Placing Performance', *Isis*, 101 (2010), 775–778.
9. Carina Bartleet and Kirsten Shepherd-Barr, 'New Directions in Theatre and Science: Guest Editorial', *Interdisciplinary Science Reviews*, 38.4 (2013), 292–294.
10. Sadiah Qureshi, *Peoples on Parade: Exhibitions, Empire and Anthropology in Nineteenth-Century Britain* (Chicago: University of Chicago Press, 2011).
11. Ervin Goffman, *The Presentation of Self in Everyday Life* (1959; London: Penguin, 1990).
12. Richard Schechner quoted in Barbara Kirshenblatt-Gimblett, 'Performance Studies'. in *The Performance Studies Reader* ed. Henry Bial (New York: Routledge, 2004), pp. 43–52 (p. 43).
13. Richard Schechner, 'The Street is the Stage', in *Performance Studies ed.* Erin Striff (Basingstoke: Palgrave Macmillan, 2003), pp. 110–123 (p. 111).
14. Erin Striff, 'Introduction: Locating Performance Studies' in P*erformance Studies* ed. Striff, pp. 1–13 (p. 11).
15. Kirshenblatt-Gimblett, 'Performance Studies', p. 50.
16. 'Wellcome Collection', Wellcome Trust, accessed via http://wellcome-collection.org/
17. 'CCIS Floor', Faculty of Science Centennial Centre for Interdisciplinary Science, University of Alberta, accessed via https://uofa.ualberta.ca/science/about-us/facilities/ccis-floor

BIBLIOGRAPHY

Bartleet, Carina and Kirsten Shepherd-Barr, 'New Directions in Theatre and Science: Guest Editorial', *Interdisciplinary Science Reviews* 38.4 (2013), pp. 292–294.

'CCIS Floor', Faculty of Science Centennial Centre for Interdisciplinary Science, University of Alberta, accessed via <https://uofa.ualberta.ca/science/about-us/facilities/ccis-floor>.

Goffman, Ervin, *The Presentation of Self in Everyday Life* [1959] (London: Penguin, 1990).

Kirshenblatt-Gimblett, Barbara, 'Performance Studies', in *The Performance Studies Reader*, ed. Henry Bial (New York: Routledge, 2004), pp. 43–52.

Latour, Bruno, *Science in Action: How to Follow Scientists and Engineers through Society* (Cambridge: Harvard University Press, 1987).

Medawar, Peter, 'Science and Literature', *Encounter*, 32.1 (1969), 15–23.

Morus, Iwan Rhys, 'Placing Performance', *Isis*, 101 (2010), 775–778.

Pickering, Andrew, *The Mangle of Practice: Time, Agency, and Science* (Chicago: University of Chicago Press, 1995).

Qureshi, Sadiah, *Peoples on Parade: Exhibitions, Empire and Anthropology in Nineteenth-Century Britain* (Chicago: University of Chicago Press, 2011).

Schechner, Richard, 'The Street is the Stage', in *Performance Studies*, ed. Erin Striff (Basingstoke: Palgrave Macmillan, 2003), pp. 110–123.

Secord, James A., 'Knowledge in Transit', *Isis*, 95 (2004), 654–672.

Shapin, Steven and Simon Schaffer, *Leviathan and the Air-Pump: Hobbes, Boyle, and the Experimental Life* (Princeton: Princeton University Press, 1985).

Shelley, Mary, *Frankenstein; Or, The Modern Prometheus* (1818; London: Penguin, 1992).

Striff, Erin, 'Introduction: Locating Performance Studies', in *Performance Studies*, ed. Striff (Basingstoke: Palgrave Macmillan, 2003), pp. 1–13.

Topham, Jonathan, 'Focus: Historicizing "Popular Science": Introduction', *Isis*, 100 (2009), 310–318.

Tyndall, John, *On the Scientific Use of the Imagination: A Discourse* (London: Longman, Green & Co., 1870).

'Wellcome Collection', Wellcome Trust, accessed via <http://wellcomecollection.org/>.

CHAPTER 2

Making the Most Beautiful Experiment: Reconstructing Gassiot's Cascade

Iwan Rhys Morus

Abstract Iwan Rhys Morus investigates the electrical performances of Charles Gassiot, who demonstrated his electrical coil, and especially its most famous instantiation as the cascade, to numerous public audiences through the 1850s. Morus considers the extensive knowledge—both technical and scientific—that was required to stage the cascade demonstrations and reveals the numerous networks of production that contributed to it. In doing this, and in revealing the complex skills required to make such performances work, Morus also shows how these popular demonstrations fed back into the skills needed for scientific experimentation and claims that without them late Victorian physics would have lacked the expertise it needed to investigate.

Over the past decade, historians of science have started to pay more attention to the visual and performative culture of Victorian science—and of science more generally.[1] This is in many ways an outcome of the turn towards practice that has typified science studies since the 1980s. Responding to the challenges of the sociology of scientific knowledge, historians started to look more carefully at what scientists did, as much as at what they said and wrote. This focus on practice meant that historians now better

I.R. Morus
University of Aberystwyth, UK

© The Editor(s) (if applicable) and The Author(s) 2016
M. Willis (ed.), *Staging Science*,
DOI 10.1057/978-1-137-49994-3_2

11

understand the role that the material and bodily culture of the sciences plays in making knowledge. As much as regarding science as an abstract and disconnected body of knowledge, historians now recognize it as a set of embodied practices. As a result, far more attention is now devoted to the places and bodies engaged in the production of scientific knowledge.[2] Whilst early pioneering exercises in dissecting scientific practice tended to focus on laboratory work and the circulation of skills between such centres of scientific production, historians more recently have started looking at public science as practice too. They have started to interrogate the ways in which the public presentation of self through different sorts of performances and technologies play a role in the construction and circulation of scientific authority.

The long nineteenth century has been a particularly fruitful area for these sorts of new approaches to understanding the public culture of the sciences for a number of reasons. One reason is the extent to which Victorian men of science participated in a wider culture of public spectacle. Much Victorian natural philosophical practice was geared towards the production of spectacle. Making nature (and the performer's own mastery over nature) visible through spectacular and practised performance was a key element of the business of natural philosophy for many of its practitioners.[3] Victorian men of science were not alone in their attempts to forge authority through such performances. Victorian public culture was also built around bodily performance as a strategy for generating and displaying authority. The proper presentation of a public self through bodily gesture and performance was central to the business of exhibiting social place, as the popularity of manuals of manners attested, amongst other things.[4] Knowing how to act and how to exhibit epistemic authority through such performances was integral to the making of a public scientific self. As a result, looking at the social and material culture of scientific performance can be highly productive as a method of understanding how scientific identity was constructed for the Victorians.

In particular such research can help illuminate the ways in which the body of the practitioner continued to matter for the making of Victorian physics. I want to approach the problem of scientific performance here by focussing on one particular item of the Victorian technology of display: Gassiot's cascade. This experiment was one of the most spectacular and widely admired electrical demonstrations of the second half of the nineteenth century. Henry Noad described the cascade as 'one of the most beautiful that can be made with the Induction Coil'.[5] He was not alone in his enthusiasm or in the language that he used to describe the effect: experimenters from Michael Faraday to John Henry Pepper reproduced Gassiot's cascade and it became a standard of experimental displays and textbook demonstrations. The disappearance of Gassiot's cascade and similar spectacular performances from our histories of Victorian physics is in many ways symptomatic of the ways in which our accounts now relegate sensational science to the margins. Experiments like the cascade belonged to a tradition of experimental performances that were designed to appeal to the sensations in general and the eye in particular. Tracing the cascade's trajectory through mid-Victorian scientific and sensational culture, and the material and cultural networks that sustained its performances, is vital in re-assessing how Victorian scientific performances worked in practice. It offers a strategy for trying to understand the bodily and material technologies that underpinned performance.

I shall begin with an article in the *Philosophical Magazine* in 1854 in which the experimenter John Peter Gassiot gave an account of the cascade for the first time.[6] His description is instructive for a number of reasons. A close reading of Gassiot's account of the cascade's genealogy offers a way of understanding the experiment as a public performance, as well as indicating how it should be situated in a broader material and social context. Gassiot's text concludes with a striking description of the cascade as a visual extravaganza. The image of electrical experiment it suggests is one of a practice tuned to the generation of spectacle. What this should reveal to us is that it was in the context of a Victorian visual culture of spectacle and spectacular performance that experiments like the cascade were understood by producers and consumers.[7] Making the cascade work within that culture required the mobilization of a wide range of material resources and bodily skills. It needed copper and glass, and the skills to mould them in particular ways.

I wish also to understand the cascade's place in the history of Victorian scientific performances. To do this, as well as to reveal how it can illuminate that history, we need to think about the position it occupied in a nineteenth-century tradition of electrical visual display. The ingredients that came together to make the cascade had their own trajectories too. The collection of copper wires, iron rods, glass, and gutta-percha required to produce the cascade, and the skills to make them do that work, were instrumental to other technologies of display and analysis too. Looking at where the cascade and the culture of spectacle and performance it instantiated belongs on those trajectories should be a salutary reminder of just how important this sensational experimental practice was for Victorian physics as we more conventionally now understand it. In a range of scientific spaces, sensation and spectacle remained central to the process of knowledge-making and its reception. There was a specific discourse of spectacle that linked different practices, instruments, and performances together.[8] The skills and resources that underpinned these experimental performances underpinned other key practices too. Knowing how to manipulate a coil mattered for the discovery of the electron as much as it mattered for playing with the cascade, for example.

Some Experiments with a Ruhmkorff Coil

The sub-heading above was the rather innocuous title of the 1854 communication to the *Philosophical Magazine* in which John Peter Gassiot described the experiment that later came to be recognized as Gassiot's cascade (or fountain).[9] Innocuous the title may be, but it is nevertheless revealing. It tells us, amongst other things, what Gassiot himself thought was the significance of the sequence of experiments leading to the cascade. Gassiot's own account of what he was doing and why it was of interest is in fact a good place to start in trying to put the cascade into a wider context of performance and spectacle. The cascade was the outcome of a series of experiments that were primarily aimed at investigating the properties of a new kind of induction coil—the Ruhmkorff coil—and finding out how it behaved. A closer look at the content of the article indicates that the investigation was aimed at establishing how the induction coil performed as a generator of spectacle. The cascade experiment was the climax of a series of tests to try out the sorts of visual displays that the new device could produce. What was being tested was the instrument's capacity for generating wonder.

Induction coils had been a standard part of electricians' armoury of instruments since their invention by Nicholas Callan in the 1830s. They were a useful source of high-intensity electricity—useful as well for medical electricity and for the nascent telegraph industry.[10] They typically consisted of two coils of copper wire, one inside the other. The inner coil was of relatively thick wire in a small number of loops, whilst the outer coil was of thinner wire in a far larger number of loops. The inner coil was connected to the poles of a voltaic battery. Each time that connection was made or broken (and induction coils typically had some simple mechanism to do this automatically, usually using an electromagnet) a very intense burst of electricity would be generated in the outer coil. As the flow of electricity in the inner coil was switched on and off very rapidly, the effect was of a continuous flow of high intensity electricity in the outer coil. During the early 1850s, the German (though resident in Paris) electrical instrument-maker Heinrich Ruhmkorff designed a number of improvements to the induction coil that made it a far more powerful piece of apparatus. It was this kind of coil that Gassiot employed his experiments.[11]

Gassiot used a coil borrowed from William Robert Grove for his early experiments, and Grove and Gassiot spent much of the rest of the 1850s alternately collaborating and competing in their investigations of discharge phenomena made possible by the use of the high-intensity electricity developed by the Ruhmkorff apparatus. Gassiot soon acquired a coil of his own, however, from the 'celebrated mechanician of Paris', as he called Ruhmkorff. His communication to the *Philosophical Magazine* described a series of experiments in which he assessed the coil's capabilities, using a battery of three Grove nitric acid cells. The coil generated a spark half an inch long through air and up to 2 inches long when passed through the flame from a spirit lamp. When the wire terminals were held a tenth of an inch apart a rapid and continuous discharge took place and the ends of the negative wire became red hot. When the discharge took place inside the vacuum of an air-pump, 'the lower half of the negative ball is surrounded by a bright, blue glow, whilst from the positive proceeds a clear, bright, red stream of light'. When a gold leaf electroscope was connected to the coil 'the discharge took place with a loud snap, the air between the plates being charged and discharged as a Leyden jar'. In effect, Gassiot was describing a process of putting the coil through its paces—seeing how it performed in producing a set of relatively common and straightforward demonstrations.[12]

The cascade itself was the culmination of this series of tests. Gassiot described the effect and it is worth quoting his words at some length:

> I coated about two-thirds of the inside of a Berlin glass beaker of 4 inches depth by 2 inches width with tin-foil, leaving about 1.5 of an inch of the upper portion uncoated. On the plate of the air-pump I placed a glass plate, and on it the glass beaker, covering the whole with an open-mouth glass receiver,

on which was placed a brass plate having a thick wire passing through collars of leather; the portion of this wire within the receiver was enclosed in an open glass tube. One end of the secondary coil was attached to the wire and the other to the plate of the air-pump. As the vacuum improves, the effect is truly surprising; at first a faint, clear, blue light appears to proceed from the lower part of the beaker to the plate; this gradually becomes brighter, until by slow degrees it rises, increasing in brilliancy until it arrives at that part which is opposite or on a line with the inner coating; the whole being intensely illuminated, a discharge then commences from the inside of the beaker to the plate of the pump in minute but diffused streams of blue light; continuing the exhaustion, at last a discharge takes place in the form of an undivided continuous stream overlapping the vessel, as if the electric fluid was itself a material body running over. When first witnessed it appears at the moment impossible to divest the mind of such a conclusion.[13]

He went on to describe the effect produced when the glass beaker was inverted so that the open end stood on the plate of the air-pump. In the next experiment a 'thin piece of talc or very thin glass coated on one side with tin-foil, and the other having a figure as a star, cross letters &c., also of tin-foil, produces a very beautiful experiment'. Finally, Gassiot revealed how the coil produced a discharge in an evacuated tube of glass, 'illuminating the whole tube'.[14]

The sequence of experiments about which Gassiot wrote were represented as experiments on and with the coil itself. Their purpose was to assess and display just what the coil—'this really beautiful instrument'—could do. The coil, in other words, was the focus of attention in all these experiments, as indeed the title of Gassiot's communication makes clear—and that was the case for the cascade experiment as much as any of the other tests of the coil's capabilities. As far as Gassiot was concerned at this stage, the cascade experiment was just another way of showing what Ruhmkorff's induction coil could do. This was not an experiment with the discharge, it was an experiment on the coil. In addition, the sequence of experiments that Gassiot related here have, in many ways, the characteristics of performance. They started with relatively simple and straightforward procedures—producing a spark, demonstrating how the presence of a flame allowed for a longer spark, demonstrating the spark in a vacuum, and so on—before going on to elicit more and more complex phenomena, culminating in the spectacle of the cascade itself. Producing spectacle, and assessing the Ruhmkorff coil's capacities as a reliable producer of spectacle, was the point of the experiment. Spectacle was not an epiphenomenon, tangential to the process of experiment, or even a fortuitous offshoot; it was what the experiment was for.

THE CULTURE OF SPECTACLE

Discharge experiments like Gassiot's cascade were part of an electrical technology of display that stretched back to eighteenth-century demonstrations such as the aurora borealis, in which the air inside a glass globe was made luminescent when it was connected to an electrical machine, or Georg Bose's notorious beatification experiments, in which an experimental subject was provided with a personal electrical halo.[15] Similar innovations elsewhere in Europe during the 1850s, such as the German instrument-maker Heinrich Geissler's development of his eponymous tubes, can also be understood as deployments of new technologies in the service of an ongoing tradition of visual practice. These technologies of display were products of a natural philosophy that operated through the senses. Conventional histories of Victorian physics sometimes chart the decline of this tradition and its replacement by a more disembodied set of practices devoted to metrology, but the popularity of experiments like the cascade should alert us to its continued flourishing into the second half of the nineteenth century.[16] If we want to understand this tradition and its appeal we need to understand its location within a wider and pervasive culture of spectacle that operated through an appeal to the senses and generated assent through visual performance. The key to forging authority in this context was display.

By the beginning of the Victorian period the philosophy of sensation was routinely invoked by natural philosophers as the basis of their claims to epistemic authority. Bacon was cited as a matter of course as the originator of a new approach to natural philosophy that received its authority from the senses. John Herschel's *Preliminary Discourse on the Study of Natural Philosophy*, for example, was very much in this tradition, and widely read amongst the gentlemen of science. 'Sensible impressions' were signals that were conveyed from external objects 'by a wonderful, and, to us, inexplicable mechanism, to our minds, which receives and reviews them [...] just as a person writing down and comparing the signals of a telegraph might interpret their meaning'.[17] To David Brewster, the eye was 'the most remarkable and the most important' of the sensory organs because of its 'boundless range of observation'.[18] Thomas Brown described the sense of vision as the key mediator between the individual and the universe. When we open our eyes, he suggested, 'it is not a small expanse of light which we perceive, equal merely to the surface of the narrow expansion of the optic nerve. It is the universe itself'.[19]

Another Scottish philosopher Thomas Reid described sight as 'without doubt the noblest' of the senses. He imagined an 'order of beings, endued with every human faculty but that of sight, how incredible would it appear to such beings, accustomed only to the flow of information of touch, that, the addition of an organ, consisting of a ball and socket of an inch diameter, they might be enabled in an instant of time, without changing their place, to perceive the disposition of a whole army, or the order of a battle, the figure of a magnificent palace, or all the variety of a landscape?'[20] According to Thomas Reid, light itself did not convey any information. What it did was provide a language of signs that could then be interpreted by the seer, so that seeing, and judging what was being seen, were both elements of the same process. One learned how to see properly. According to Thomas Brown, it was 'the mixed product of innumerable observations, and calculations, and detections of former mistakes, which were the philosophy of our infancy, and each of which, separately, has been long forgotten, recurring to the mind, in after-life, with the rapidity of an instinct'.[21] So seeing was the model for knowing—as Reid put it, the 'evidence of reason is called *seeing*, not *feeling, smelling*, or *tasting*'.[22]

These sorts of philosophies of visual sensation in turn fed into a prevailing culture of popular and sensational spectacle that played quite deliberately and subversively with this sense of the visual as authoritative. Panoramas and phantasmagoria offered audiences fully immersive sensory experiences that worked at convincing viewers of the reality of the spectacles they witnessed by manipulating the eye. Spectacles like these operated by a careful spatial choreography that directed vision in particular ways, as Willis also argues in this volume in his analysis of travel guides. Marvelling at the London Colosseum's panorama of the city as seen from the summit of St. Paul's cathedral dome, one commentator suggested that the viewer was 'obliged almost to reason with yourself, to be persuaded that it is not nature, instead of a work of art, upon which you are bestowing your admiration'.[23] The comment summed up the attraction of many of these sorts of spectacles for their spectators. They operated in the grey area between the natural and the artificial and challenged their viewers to make the discrimination. Phantasmagoria were just the same: their audiences had to decide whether the ghosts they saw were real or carefully contrived artefacts—whilst giving way to pleasurable thrills of terror at the same time. Spectacles like these both depended on the authority of vision and raised questions about its reliability.

Visual deception was an important component of this culture of spectacle. Both panorama and phantasmagoria worked through fooling the eye. They were performances that made the instability of vision part of the spectacle itself. Natural philosophers like Brewster or Herschel, whilst extolling the eye as the most authoritative foundation for knowledge about the world, also emphasized its fallibility. Herschel lectured his readers on the importance of combating what he called 'prejudices of sense' through proper and disciplined observation and reasoning. 'Not to trust the evidence of our senses, seems, indeed, a hard condition', he admitted, but 'it is not the direct evidence of our senses that we are called upon to reject, but only the erroneous judgements we unconsciously form from them'.[24] Brewster was less sanguine about the power of judgement. Discussing the difficulty in distinguishing between embossed and depressed images (cameos and intaglios), he remarked that the 'greater our knowledge is, of the subject, the more readily does the illusion seize upon us; while, if we are but imperfectly acquainted with the effects of light and shadow, the more difficult is it to be deceived'.[25] His *Letters on Natural Magic* paid particular attention to illusions like these precisely because they could be so revealing about the fallibility of what remained nevertheless the best mechanism for knowing.

The creation of artful illusions of reality was an important feature of theatre practice. Theatre managers wanted what their audiences saw on stage to appear as convincing as possible a simulacrum of nature. Throughout the nineteenth century, theatres competed to mount performances using complex stage machinery to generate wonder for their audiences. Some critics were unimpressed. G.H. Lewes complained in 1853 about audiences 'who think scenery and costume the "be-all and the end-all of the drama"' and that 'Drama' had become 'nothing more than a Magic Lantern on a large scale'.[26] When Pepper's ghost appeared in theatrical productions outside its original home in the Royal Polytechnic Institution, some critics complained that this was 'a startling addition to stage machinery, and a terrible extension of a new style of attraction which has sprung up during the last ten years.' They argued that the 'attractions, as well as the merits, of a performance should depend as much as possible upon the actors—as little as possible upon accessories which properly belong to other branches of art'.[27] In some performances, such as the ballet *Electra, or the Lost Pleiade* mounted at Her Majesty's Theatre in

1849, the spectacle generated by the technology of display—in this case the electric arc light—provided the rationale for the whole performance.[28]

Electric arc lights (and Pepper's ghost for that matter) offer striking examples of just how fluid and permeable the boundary between science and theatre was in this culture of spectacle. At venues like the Adelaide Gallery, the Royal Polytechnic Institution, the Royal Panopticon of Science and Art, or even the Department of Natural Magic at the Colosseum, complex and sophisticated performances of spectacular experimental displays were the major attraction. Occasions like the unveiling of the Polytechnic's hydro-electric machine in 1843 or Pepper's gargantuan Monster Coil in 1869 provided opportunities for displays of wonder. With the hydro-electric machine, the 'passage of the electricity over the tinfoil on the tubes was far more brilliant, and the aurora borealis exceeded in intensity and beauty anything we had ever witnessed'. The Monster Coil's dimensions—150 miles of wire in the secondary coil and a primary coil that weighed 145 pounds—were a source of spectacle in themselves, even before the machine was put into action. The vision of science that spaces like these offered was one in which experiment was entirely geared towards the production of spectacle. Generating spectacle and generating knowledge were the same thing: 'every marked increase in the power of scientific apparatus has been followed by a corresponding increase in the growth of knowledge', commented the *Times* of the Monster Coil.[29]

Accounts of the spectacles generated by these technologies of display suggest that there was a specific aesthetic being appealed to in these performances. Looking at the evanescent, transient glows produced by these instruments was clearly meant to be an aesthetic experience, like looking at a marvellous sunset, or a Turner painting. Audiences went to places like the Adelaide Gallery expecting to be entertained and stimulated by what they saw in a particular way. Encountering science in such spaces was an education for the eye. Scientific spectacles like Gassiot's cascade and other examples of the tradition of visual practice to which the experiment belonged made nature visible, but also highlighted the artefactual nature of knowledge.[30] They made the work (and authority) of the performer visible too. Getting the spectacle right was certainly neither artless nor effortless. When the Polytechnic's hydro-electric machine was first unveiled to the public, newspapers noted that the demonstrator George Bachhoffner had spent a day beforehand practising his performance.[31] Making the spectacle seamless needed careful choreography. Moreover, the spaces where the culture of spectacle was made visible, and the machinery through which it was performed, were themselves parts of wider networks of production, distribution, and consumption.

MATERIALITY AND THE TECHNOLOGY OF DISPLAY

Paying attention to these wider networks is essential if we want to understand the conditions that needed to be in place in order to make experimental performances like the cascade possible. The experimental culture that generated the cascade depended on the existence of skilled individuals and material resources for its success. Those skills and resources themselves in turn did not exist in a vacuum. They were the products of quite specific historical circumstances. So if we want to understand what the cascade was made of, and how it was made to work, we need to look at just where particular material resources came from and how they were organized, where the skills of manipulating those resources had their origins and how they circulated. Experiments like the cascade were the outcome of organized work. They were founded on the production of raw material and its distribution. They required the skills of metalworkers, glassworkers, and instrument-makers. They depended on the intimate knowledge of how particular configurations of apparatus might respond to manipulation that came with long practice and with membership of a core set of practitioners. Turning that collective tacit knowledge into the public performance called Gassiot's cascade was an achievement in the marshalling and deployment of material and cultural resources.

We can trace the origins of the collection of instrumental practices and techniques that led to the induction coil back to William Sturgeon's discovery in 1824 that it was possible to turn a soft iron core into a magnet by passing electricity through a coil wound about it. The resulting electromagnet was one of a suite of table-top instruments that Sturgeon presented to the Society of Arts to be awarded 30 guineas and a silver medal.[32] It was a collection of instruments designed with spectacle in mind. Sturgeon and others such as the American electrician Joseph Henry and the Dutch philosopher Gerrit Moll experimented with different ways of winding the coils that led to huge increases in magnetic power. Henry himself noted that one particular advantage that the electromagnet possessed as a tool for investigating the relationship of magnetism to electricity was that the magnetism could be switched on and off or reversed at will: 'their polarity can be instantaneously reversed, and their magnetism suddenly destroyed or called into full action, according as the occasion may require'.[33] By finding more efficient ways of winding the coils Henry succeeded in building electromagnets that could carry as much as 1600 pounds·

Investigations like these were given new impetus in 1831 with Michael Faraday's discovery of electromagnetic induction. Faraday's laboratory notebook simply noted that he had 'had an iron ring made (soft iron) [...]. Wound many coils of copper wire round one half, the coils being separated by twine and calico [...]. On the other side but separated by an interval was wound wire in two pieces together amounting to about 60 feet in length'.[34] It seems likely that he had been investigating in a similar way to Sturgeon, Moll, and Henry with coils and iron cores. In the aftermath of Faraday's discovery, experimenters, including Faraday himself, competed to find ways of making the effect into a spectacle. That was the immediate context for Nicholas Callan's invention of the induction coil in 1836. The coil (which, interestingly, he initially described as an 'electromagnet') was the outcome of improvisational experiments to try and make electromagnetic effects more spectacular. 'From all the experiments which I have made on the magneto-electric shock', he suggested, 'I think I may fairly conclude, that, if 2000 feet of wire were coiled on a bar of soft iron 6 feet long and an inch thick, a shock might be obtained with the aid of a single pair of plates, which would equal that of a battery of 100 voltaic circles'.[35]

Callan's invention was one of many mid-1830s' contrivances to exploit the new electromagnetic phenomenon that Faraday had identified. The issue of the *Philosophical Magazine* in which he published his announcement contained accounts of other similar efforts, such as Joseph Saxton's magneto-electric device. Saxton's apparatus had been developed as part of his duties as the Adelaide Gallery's resident instrument-maker. There he was expected to keep up a flow of novel inventions and spectacles that would continue to entice the public through the gallery's door. The magneto-electric apparatus was itself an improvement of an earlier device that used Faraday's discovery to produce a spark from an electromagnet. The diary that Saxton kept during this period offers an interesting view of the networks of production and consumption to which the Adelaide Gallery belonged and through which Saxton moved to gather materials and resources for his displays.[36] Instruments like the induction coil and the magneto-electric machine were designed to produce spectacle—and they were assessed through their capacity for such performances. On 20 June 1833, for example, Saxton recorded in his diary: 'Made a tryal [sic] of the new arrangement of the Magneto-Electrical Machine, found it to surpass expectations. It produces a continued spark and so mutch a shock to the tong and lips that it is impossible to bear it for any length of time'.[37]

The experiments with sparks, and later with electrical discharges in vacuum tubes, that Grove and Gassiot engaged in during the 1840s and that led to the cascade, were part of this tradition of experimental spectacle and depended on these sorts of technologies. Grove had been experimenting with sparks since 1840, when he had carried out investigations of the relationship between the colour of sparks and the kinds of metal used to produce them. This was all in the context of his broader concern with building better batteries that had led to his invention of the nitric acid battery a few years earlier, as well as leading to the gas battery a year later in 1841.[38] By the 1850s both Grove and Gassiot were interested in investigating the brightly coloured discharges that appeared between the poles of high-intensity devices like the induction coil when enclosed in glass tubes that had been evacuated (or partially evacuated) of air. It was for this kind of experimental work that Grove purchased the Ruhmkorff coil that he later lent to Gassiot for his experiments.[39] Work like this needed equipment that could operate reliably at high electrical tensions. In other words it, depended on the availability of highly skilled instrument-makers with access to the right resources.

An increasingly important factor in the availability of these skills and resources was the growing telegraph industry. Telegraphy was itself a product of exactly this culture and technology of display. Making instruments that worked effectively over long distances needed exactly the skills and knowledge that experimenters working with coils and electromagnets had amassed whilst looking for new ways to exhibit spectacular phenomena. Samuel Morse had considered that if 'the presence of electricity can be made visible in any part of the circuit, I see no reason why intelligence may not be transmitted by electricity'.[40] In practice it needed the skills of coil winding that Charles Wheatstone had learned from the electromagnet-maker Joseph Henry to do that over any distance. As the telegraph industry grew during the second half of the 1840s it both depended on and helped create a pool of exactly those kinds of skills. It also created a market for reliable electrical instrumentation. One of the features of the Ruhmkorff coil to which Grove drew particular attention, for example, was 'the attention paid to insulation in the construction of this apparatus'.[41] Effective insulation was also a prerequisite of long-distance telegraphy.

In its most fundamental aspects, Gassiot's cascade depended on copper, glass, and gutta-percha. Those were the main components of the instruments used for its performance. The cascade was at one end of a network of connections that included copper mines in Anglesey, Cornwall, or Devon, copper-smelting centres like the Swansea valley, and wire-drawing factories in London and elsewhere. The main users of copper wire of any kind before the rise of the telegraph industry during the 1840s had been jewellers and toy-makers—both crafts that had some close connections with philosophical instrument-makers. Insulation was another requirement. Electricians often used strips of silk or a similar thin fabric for this purpose—and again the process of winding this material around the wire was both tedious and time-consuming. From the 1850s onwards gutta-percha was increasingly commonly used as a form of insulation for electrical wires—primarily in the telegraph industry. Induction coils such as the one Gassiot used needed insulation not only for the wires but at the ends of the coil and on components such as the switch which would otherwise be difficult to manipulate whilst the coil was in action. Hence, rubber cultivators on the Malay Peninsula also had a role to play in providing the resources that made the cascade possible. Another prerequisite was high quality glass. Glass-making skills were therefore also part of the complex networks that underpinned the experiment. London itself was a centre for glass-making. The best sand for the process came from the beaches of Lynn and the Isle of Wight, and the highest quality potash came from Canada.[42]

What I want to highlight here is the extent to which experimental performances like Gassiot's cascade were comprehensively embedded in the industrial-imperial networks of early Victorian Europe. The material culture and the networks of skills and resources needed in order to perform such experiments were not just matters of science. They tied early Victorian experimental culture and its performances to the material conditions that made them possible. The raw materials that constituted the cascade had to be mined and harvested, they had to be carried around, they had to be shaped and finished before they could become elements in an induction coil, an air-pump, or even a wine glass. The practical skills that underpinned the cascade's performances also fed back into the networks that sustained it. They were entangled with the skills that underpinned a range of industrial performances. When performers at the Royal Polytechnic Institution or elsewhere mounted a demonstration of Gassiot's cascade they were participating not just in the mid-Victorian culture of scientific spectacle, but in industrial and imperial culture too. The experiment they were showing off was as much a product of empire and industry as was a steam engine in a Victorian cotton mill.

CONTEXTS AND TRAJECTORIES OF PERFORMANCE

Gassiot's cascade sits at an interesting point of intersection between different and entangled experimental trajectories during the 1850s. This is one of the reasons it makes such a fascinating example of Victorian scientific performance. The cascade certainly became a staple item in mid- and later Victorian technologies of spectacular experimental display. The popular lecturer and author Henry Noad, as we have seen, described the cascade as 'one of the most beautiful that can be made with the Induction Coil' and a 'truly magnificent experiment' (and prescribed Geissler's mercury air-pump as the best means of producing the required vacuum).[43] John Henry Pepper commented that comparing 'so many beautiful experiments, it is somewhat difficult to say which is the most pleasing, but for softness and exquisite colouring, with the continuous vibrating motion of the flowing current of electricity, nothing can surpass the "cascade experiment"'.[44] Like Noad, Pepper offered his readers careful instructions on how to reproduce the effect so that 'a continuous series of streams of electric light seem to overflow the goblet all round the edge, and it stands then the very embodiment of the brimming cup of *fire*, and emblematical of the dangers of the wine-cup'.[45] It was one of the most spectacular items in what seems to have been a standard repertoire of experimental performances to show off the capacities of an induction coil.

Some descriptions of the cascade offered variations on the basic theme. Noad, for example, suggested modifications to the original experiment that would make it simpler to perform. Later in the century it was common to use a wine glass made of uranium glass so that the glass itself glowed green as well. It also seems clear that the experiment was to some extent being reduced into a standard package. One popular turn of the century text described the apparatus needed for 'one of the most beautiful experiments that can be performed' as

> a cylinder of glass, slightly enlarged at each end, bulb form. The upper bulb is divided from the rest of the cylinder by a glass partition, in the centre of which is a tube reaching down towards the lower bulb. This, like the upper, is separated by a glass partition, having one or more holes through it. On this lower partition is fixed a glass goblet, so arranged that the tube from the upper bulb comes down inside the goblet nearly to its bottom. The platinum wires for the electrodes are fixed one in the upper and the other in the lower bulb.[46]

The complex performance had been replaced by a standardized piece of equipment.

The transformation should remind us of the cascade's position in wider cultures of electrical performance. It was one of a wide array of discharge demonstrations involving vacuum tubes that became popular from the 1850s onwards. Whilst the cascade was certainly amongst the most complex of these performances, it shared a number of characteristics with other experiments of the genre. More or less contemporaneous with the cascade were the eponymous Geissler tubes invented by the German instrument-maker Heinrich Geissler. These were enclosed glass tubes, usually twisted into intricate shapes with an electrode at each end. They were filled with different luminescent gases and glowed brightly when electricity from a coil was passed through them. Gassiot used Geissler's tubes in some of his own discharge experiments later in the 1850s. In one of the series of communications on discharge phenomena he addressed to the Royal Society, Gassiot described how he 'had the opportunity of experimenting with upwards of sixty of Geissler's vacua-tubes, in which many beautiful and novel results are produced'.[47] Experiments like these (and contemporaneous ones carried out by Grove and others) were simultaneously demonstrations of the discharge's character that appealed to the sensations and investigations into its nature. Grove in his 1852 communication on the electrochemical polarity of gases described an experiment he had carried out with Gassiot:

> A cone of blue flame was now perceptible, the water forming its base, and the point of the wire its apex; the wire rapidly fused, and became so brilliant that the cone of flame could be no longer perceived, and the globule of fused platinum was apparently suspended in air and hanging from the wire; it appeared sustained by a repulsive action, like a cork ball on a jet d'eau, and threw out scintillations in a direction away from the water.[48]

There was clearly a sensuous aesthetic of experimentation at play here and that aesthetic can be followed through subsequent experiments on discharge phenomena during the following decades. Warren de la Rue and Cromwell Varley's experiments on discharge phenomena during the 1860s and 1870s were in the same tradition of visual practice—as were William Crookes's experiments on the 'fourth state of matter' at about the same time. Varley offered striking descriptions of the effects generated by his experiments: 'a tongue of light projected from the positive pole towards the negative, the latter being still completely obscure. The light around the positive pole was to our eyes white, while the projecting flame was a bright brick-red'.[49] Varley experimented with using photography to capture these wonder-inducing, if highly transient and liminal phenomena. Crookes was fascinated by the radiant matter (his 'fourth state') flowing between the poles in the discharge tube. In one of his most striking demonstrations of this fourth state of matter and its properties he arranged a tiny glass locomotive driven by miniature paddles on a railway track inside a discharge tube. The flowing streams of radiant matter between the poles made the paddle wheels rotate, driving the locomotive along the track and providing visual evidence of electrical power to demonstration audiences.

The mathematical physicist George Gabriel Stokes was an admirer of Crookes's flair as an experimental performer, if not of his acumen as a theoretician. He wrote to a colleague that for 'enlarging our conceptions of the ultimate structure of matter, I know of nothing like what Crookes has been doing for some years. I wish you could see some of the work in his laboratory.' The remark should remind us that these experimental performances belonged in the world of metrological exactitude as much as they were part of the culture of spectacle. For example, the apparatus that J.J. Thomson, the Cavendish professor of experimental physics at Cambridge, used during the series of experiments during the 1890s that led to the discovery of the electron reveals a material culture not very different from that of Gassiot's cascade. The glass tubes, vacuum pumps, induction coils, and batteries that were used to discover the electron belonged to the same tradition of experimental practice as the equipment that Gassiot deployed to make the cascade 43 years earlier. Thomson's experiments were in a straight line of descent from those conducted by Gassiot or Grove during the 1850s and depended on the same sorts of networks of cultural and material resources.[50]

This line of descent is one reason, at least, why these spectacular demonstrations cannot be dismissed straightforwardly as marginal to the mainstream of Victorian physics, even if they have largely disappeared from our conventional histories of the period's scientific achievements. It was in the context of developing experimental performances such as Gassiot's cascade that the material resources and instrumental skills that led to the foundational experiments of atomic physics half a century later were first developed. Without the culture of spectacle within which the cascade was created there would have been no tradition of skills and resources that later workers might draw upon. If we want to understand how atomic physics originated, we therefore need to understand experimental performances like these and their contexts. More importantly, though, the positions occupied by experimental performances like Gassiot's cascade in the entangled trajectories of mid-Victorian physics should draw our attention to the open-endedness and interpretative flexibility of those performances. For experimental performers like Gassiot, and others in his tradition of visual manipulation, generating spectacle was as much a part of the process of physics as was the interrogation of nature—or rather interrogating nature and making nature into a spectacle were the same thing.

My concern in this chapter has been to explore the materiality of scientific performances like Gassiot's cascade and by doing that to look at the place of such experiments in the Victorian culture of spectacle. Contemporary replication is one method that helps to illuminate the perfomative elements of such demonstrations. Reconstructing the experiment of Gassiot's cascade is a methodologically interesting and fruitful way of getting to grips with the material requirements of such performances and the skills required to make them work.[51] Reconstruction revealed, for example, how much materiality mattered. The shape of the glass turned out to have a significant role to play in getting the phenomenon right. Performing the experiment was also a good way of thinking about the reception of such performances—how were they seen (literally) by their audiences and what kind of choreography needed to be deployed to make the performance effective as a piece of showmanship.

Thinking about the material and bodily culture of performance as embodied in experiments like the cascade brings home the important place of sensation in the culture of Victorian physics. The most striking thing about the cascade is (and was) its tangibility. This was an experiment designed to create sensation. It was also an experiment that worked by fooling the eye. It appeared to the viewer as though it were a stream of fire but it was clearly designated within the performance as something other than that. It simultaneously appealed to the eye whilst reminding the spectator that what they saw was suspect. The cascade was sensational because it made the immaterial and liminal appear concrete and material, and by doing so obliged its spectators to think again about what they were seeing. The cascade's appeal to sensation and the way in which it played with the senses should make us think again about the direction of travel of Victorian physics. We have seen that Gassiot's cascade straddled a number of intersecting traditions within Victorian physics. It shared material culture and practice with telegraphy and discharge tube physics as well as the world of scientific spectacle. The cascade demonstrates that spectacle and sensation remained important as strategies for generating knowledge throughout the Victorian period.

NOTES

1. Aileen Fyfe and Bernard Lightman, eds *Science in the Marketplace: Nineteenth-Century Sites and Experiences* (Chicago: University of Chicago Press, 2007). See also the *Isis* Focus section edited by Iwan Rhys Morus: 'Performing Science', *Isis*, 101 (2010), 775–828.
2. Christopher Lawrence and Steven Shapin, eds *Science Incarnate: Historical Embodiments of Natural Knowledge* (Chicago: University of Chicago Press, 1998); David Livingstone, *Putting Science in Its Place: Geographies of Scientific Knowledge* (Chicago: University of Chicago Press, 2003).
3. Bernard Lightman, *Victorian Popularizers of Science: Designing Nature for New Audiences* (Chicago: University of Chicago Press, 2007).
4. Leonore Davidoff, *The Best Circles* (London: Ebury Press, 1986); Andrew St. George, *The Descent of Manners: Etiquette, Rules and the Victorians* (London: Chatto & Windus, 1993).
5. Henry Minchin Noad, *The Inductorium* (London: John Churchill and Sons, 1866), p. 59.
6. John Peter Gassiot, 'On Some Experiments with Ruhmkorff's Induction Coil', *Philosophical Magazine*, 7 (1854), 97–99 (p. 99).
7. Iwan Rhys Morus, *Frankenstein's Children: Electricity, Exhibition and Experiment in early Nineteenth-century London* (Princeton: Princeton University Press, 1998).
8. Iwan Rhys Morus, 'Illuminating Illusions, or, the Victorian Art of Seeing Things', *Early Popular Visual Culture*, 10 (2012), 37–50. Such discourses already had a long history: see Simon Schaffer, 'Natural Philosophy and Public Spectacle in the Eighteenth Century', *History of Science*, 21 (1983), 1–43.
9. Gassiot, 'Induction Coil'.
10. Morus, *Frankenstein's Children*; Brian Gee, 'The Early Development of the Magneto-electric Machine', *Annals of Science*, 50 (1993), 101–133.
11. Iwan Rhys Morus, *When Physics Became King* (Chicago: University of Chicago Press, 2005), p. 105.
12. Gassiot, 'Induction Coil', p. 98.
13. Gassiot, 'Induction Coil', p. 99.
14. Gassiot, 'Induction Coil', p. 99.
15. John Heilbron, *Electricity in the 17th and 18th Centuries: A Study in Early Modern Physics* (Berkeley and Los Angeles: University of California Press, 1979).

16. Robert D. Purrington, *Physics in the Nineteenth Century* (New Brunswick: Rutgers University Press, 1997); Peter Harman, *Energy, Force and Matter: The Conceptual Development of Nineteenth-century Physics* (Cambridge: Cambridge University Press, 1982).

17. John Herschel, *Preliminary Discourse on the Study of Natural Philosophy* (London: Longman, Rees, Orme, Brown, and Green, 1831), p. 84.

18. David Brewster, *Letters on Natural Magic* (London: John Murray, 1832), p. 8.

19. Thomas Brown, *Lectures on the Philosophy of the Human Mind* (Edinburgh: William Tait, 1840), p. 178.

20. Thomas Reid, *An Inquiry into the Human Mind* (Edinburgh: Bell and Bradfute, 1801), p. 154.

21. Brown, *Lectures*, p. 181.

22. Reid, *Inquiry*, p. 155.

23. Quoted in Richard D. Altick, *The Shows of London* (Cambridge MA: Belknap Press, 1978), p. 149.

24. Herschel, *Discourse*, p. 81.

25. David Brewster, 'On the Optical Illusion of the Conversion of Cameos into Intaglios, and of Intaglios into Cameos, with an Account of other Analagous Phenomena,' *Edinburgh Journal of Science*, 4 (1826), 99–108 (p. 104).

26. Quoted in Altick, *Shows*, p. 186.

27. Anon., 'Public Amusements', *Lloyd's Weekly Newspaper*, 28 June 1863, n.p.

28. This is particularly interesting in the light of Shepherd-Barr's later chapter in this volume, which argues that contemporary science plays are, similarly, focussed as much on science as they are on dramaturgy (see Chap. 5).

29. Quotes from Iwan Rhys Morus, 'More the Aspect of Magic than Anything Natural: The Philosophy of Demonstration in Victorian Popular Science,' in Fyfe and Lightman, eds *Science in the Marketplace*, pp. 336–370.

30. Morus, *Frankenstein's Children*.

31. Morus, 'Magic'.

32. Morus, *Frankenstein's Children*.

33. Jospeh Henry, 'On the Production of Currents and Sparks of Electricity from Magnetism', *American Journal of Science*, 22 (1832), 403–408 (p. 403).

34. Thomas Martin, ed *Faraday's Diary* 7 vols (London: G. Bell & Sons, 1932), vol. 1, p. 367.

35. Nicholas Callan, 'On a New Galvanic Battery', *Philosophical Magazine*, 9 (1836), pp. 472–478, 476.

36. Morus, *Frankenstein's Children*.

37. Quoted in Morus, *Frankenstein's Children*, p. 87.

38. Iwan Rhys Morus, *William Robert Grove: Victorian Gentleman of Science* (Cardiff: University of Wales Press, 2016).
39. William Robert Grove, 'On the Electro-chemical Polarity of Gases', *Philosophical Magazine*, 4 (1852), pp. 498–515.
40. Edward Lind Morse, *Samuel F. B. Morse—His Letters and Journals*, 2 vols. (Boston: Houghton Mifflin, 1914), vol. 2, p. 6.
41. Grove, 'Polarity', p. 501.
42. George Dodds, 'A Day at a Flint-glass Factory', in *Days at the Factories* (London: Charles Knight, 1843), pp. 257–280.
43. Noad, *Inductorium*, p. 59.
44. John Henry Pepper, *The Boy's Playbook of Science* (London: George Routledge and Sons, 1869), p. 268.
45. Pepper, *Playbook*, p. 269.
46. Anon, *Intensity Coils: How Made and Used* (London: Perken, Son and Co., 1901), p. 53.
47. John Peter Gassiot, 'On the Stratification in Electrical Discharge', *Philosophical Transactions*, 149 (1859), 137.
48. Grove, 'Polarity', p. 500.
49. Cromwell Varley, 'Some Experiments on the Discharge of Electricity though Rarified Media and the Atmosphere', *Proceedings of the Royal Society*, 19 (1871), 236.
50. Jaume Navarro, *A History of the Electron: J. J. and G. P. Thomson* (Cambridge: Cambridge University Press, 2012).
51. For discussions of replication as historical strategies see Heinz Otto Sibum, 'Reworking the Mechanical Value of Heat: Instruments of Precision and Gestures of Accuracy in Early Victorian England', *Studies in History & Philosophy of Science*, 26 (1995), 73–106; Peter Heering, 'Regular Twists: Replicating Coulomb's Wire-Torsion Experiments', *Physics in Perspective*, 8 (2006), 52–63.

BIBLIOGRAPHY

Altick, Richard D., *The Shows of London* (Cambridge MA: Belknap Press, 1978).
Anon., 'Public Amusements', *Lloyd's Weekly Newspaper*, 28 June 1863, unpaginated.
Anon., *Intensity Coils: How Made and Used* (London: Perken, Son & Co., 1901).
Brewster, David, 'On the Optical Illusion of the Conversion of Cameos into Intaglios, and of Intaglios into Cameos, with an Account of other Analagous Phenomena,' *Edinburgh Journal of Science*, 4 (1826), 99–108.
Brewster, David, *Letters on Natural Magic* (London: John Murray, 1832).
Brown, Thomas, *Lectures on the Philosophy of the Human Mind* (Edinburgh: William Tait, 1840).

Callan, Nicholas, 'On a New Galvanic Battery', *Philosophical Magazine*, 9 (1836), 472–78.

Davidoff, Leonore, *The Best Circles* (London: Ebury Press, 1986).

Dodds, George, 'A Day at a Flint-glass Factory', *Days at the Factories* (London: Charles Knight, 1843), pp. 257–80.

Fyfe, Aileen and Bernard Lightman, eds *Science in the Marketplace: Nineteenth-Century Sites and Experiences* (Chicago: University of Chicago Press, 2007).

Gassiot, John Peter, 'On Some Experiments with Ruhmkorff's Induction Coil', *Philosophical Magazine*, 7 (1854), 97–99.

Gassiot, John Peter, 'On the Stratification in Electrical Discharge', *Philosophical Transactions*, 149 (1859), 137.

Gee, Brian, 'The Early Development of the Magneto-electric Machine', *Annals of Science*, 50 (1993), 101–33.

Grove, William Robert, 'On the Electro-chemical Polarity of Gases', *Philosophical Magazine*, 4 (1852), 498–515.

Harman, Peter, *Energy, Force and Matter: The Conceptual Development of Nineteenth-century Physics* (Cambridge: Cambridge University Press, 1982).

Heering, Peter, 'Regular Twists: Replicating Coulomb's Wire-Torsion Experiments', *Physics in Perspective*, 8 (2006), 52–63.

Heilbron, John, *Electricity in the 17th and 18th Centuries: A Study in Early Modern Physics* (Berkeley and Los Angeles: University of California Press, 1979).

Henry, Joseph, 'On the Production of Currents and Sparks of Electricity from Magnetism', *American Journal of Science*, 22 (1832), 403–08.

Herschel. John, *Preliminary Discourse on the Study of Natural Philosophy* (London, Longman, Rees, Orme, Brown, and Green, 1831).

Lawrence, Christopher and Steven Shapin, eds *Science Incarnate: Historical Embodiments of Natural Knowledge* (Chicago: University of Chicago Press, 1998).

Lightman, Bernard, *Victorian Popularizers of Science: Designing Nature for New Audiences* (Chicago: University of Chicago Press, 2007).

Livingstone, David, *Putting Science in Its Place: Geographies of Scientific Knowledge* (Chicago: University of Chicago Press, 2003).

Martin, Thomas, ed *Faraday's Diary* 7 vols (London: G. Bell & Sons, 1932).

Morse, Edward Lind, *Samuel F. B. Morse – His Letters and Journals*, 2 vols. (Boston, Houghton Mifflin, 1914).

Morus, Iwan Rhys, *Frankenstein's Children: Electricity, Exhibition and Experiment in early Nineteenth-century London* (Princeton: Princeton University Press, 1998).

Morus, Iwan Rhys, *When Physics Became King* (Chicago: University of Chicago Press, 2005).

Morus, Iwan Rhys, 'More the Aspect of Magic than Anything Natural: The Philosophy of Demonstration in Victorian Popular Science', in *Science in the Marketplace*, eds Fyfe and Lightman, pp. 336–70.

Morus, Iwan Rhys, 'Performing Science', *Isis*, 101 (2010), 775–828.

Morus, Iwan Rhys, 'Illuminating Illusions, or, the Victorian Art of Seeing Things', *Early Popular Visual Culture*, 10 (2012), 37–50.

Morus, Iwan Rhys, *William Robert Grove: Victorian Gentleman of Science* (Cardiff: University of Wales Press, 2016).

Navarro, Jaume, *A History of the Electron: J. J. and G. P. Thomson* (Cambridge: Cambridge University Press, 2012).

Noad, Henry Minchin, *The Inductorium* (London, John Churchill & Sons, 1866).

Pepper, John Henry, *The Boy's Playbook of Science* (London, George Routledge & Sons, 1869).

Purrington, Robert D., *Physics in the Nineteenth Century* (New Brunswick: Rutgers University Press, 1997).

Reid, Thomas, *An Inquiry into the Human Mind* (Edinburgh, Bell & Bradfute, 1801).

Schaffer, Simon, 'Natural Philosophy and Public Spectacle in the Eighteenth Century', *History of Science*, 21 (1983), 1–43.

Sibum, Heinz Otto., 'Reworking the Mechanical Value of Heat: Instruments of Precision and Gestures of Accuracy in Early Victorian England', *Studies in History & Philosophy of Science*, 26 (1995), 73–106.

St. George, Andrew, *The Descent of Manners: Etiquette, Rules and the Victorians* (London: Chatto & Windus, 1993).

Varley, Cromwell, 'Some Experiments on the Discharge of Electricity though Rarified Media and the Atmosphere', *Proceedings of the Royal Society*, 19 (1871), 236.

CHAPTER 3

Science in the City: Scientific Display and Urban Performance in Victorian Travel Guides to London

Martin Willis

Abstract Willis offers the first analysis of scientific representations in travel guide books to London. Taking a range of travel guides from the 1860s to the end of the century, Willis considers which scientific sites travel guides chose to promote to the tourist reader, and investigates what kinds of representations of science their narratives provide. This allows for two key questions: first, how did science contribute to visions of London as a city, and second, in what ways did travel guides mediate science to achieve this. Willis concludes that travel guides are the location of a particularly rich interaction between science and the humanities, where literary writers (often of some repute) give science a powerful place in an emerging modernity through language.

Alighting from a train in the heart of London in 1863, a tourist who had digested her newly published *Bradshaw's Descriptive Railway Handbook* would know she had arrived in the capital of 'the civilized world' (p. 5). 'The British metropolis', *Bradshaw's* begins, 'contains the largest mass of

M. Willis
Cardiff University, UK

© The Editor(s) (if applicable) and The Author(s) 2016   35
M. Willis (ed.), *Staging Science*,
DOI 10.1057/978-1-137-49994-3_3

human life, arts, science, wealth, power, and architectural splendour [...] that ever has existed in the known annals of mankind' (p. 5). If the tourist followed the guidebook writer's advice to first 'walk through the principal streets' (p. 6), under *Bradshaw's* guidance of course, she would soon encounter the Strand, and on it Somerset House, 'a magnificent pile of buildings', perhaps 'the finest object of the kind in London' which houses 'the apartments of the Royal Society, and Society of Antiquaries' (p. 7). Indeed in walking through almost any part of London—up Regent Street from Oxford Circus, into Chelsea, or over to Greenwich—the tourist, constructed for the purposes of the guidebook as the awestruck stranger playing witness to London's spectacular displays of urban modernity, would encounter places and spaces of science. The Zoological Gardens in Regent's Park, the military Hospital, the Royal Observatory, as well as numerous other sites of scientific activity, are woven into *Bradshaw's* depiction of the contemporary cityscape. Science is enlisted, in such guidebook representations, as an actor in the dramatic presentation of the modern city. Various performances are demanded of it: science must depict the future, act out exciting experimentalism, or as in *Bradshaw's* script for Somerset House, make concrete the elite longevity of London's intellectual culture. For the tourist following the itinerary in her handy *Bradshaw's* guidebook, however, science would not look as though it were performing. The guidebook, after all, claims that it is not acting, that its performance is only that of the everyday and the informational.[1] The epistemology of the guidebook, it seems, is to make plain and not to obfuscate, to offer simple accounts of the real rather than falsely constructing the dramatic. The tourist following her *Bradshaw's* is, then, in a particularly interesting position with regard to London's science. For she is invited to read and see the science of the city as familiar and ordinary while experiencing it on the page and within the cityscape as something unique and unusual.

Examinations of science in the city remain underdeveloped, despite the turn towards thinking about science in spatial terms. As Sven Dierig, Jens Lachmund, and Andrew Mendelsohn claimed, quite rightly, in 2003, 'the city has received remarkably scant attention as the most direct physical and socio-spatial locus of science'.[2] Work since then, especially by historical geographers, has opened up both new avenues for research and new methods for pursuing it. Charles Withers and David Livingstone, in particular, argue convincingly in their work on nineteenth-century science for the centrality of thinking about 'the making and meaning of science *in place*' as well as 'science's movement *over space*'.[3] And they highlight how theorists of space—like Henri Lefebvre or Michel De Certeau—might offer valuable ways of thinking about space itself that productively complicate present assumptions and lead to new insights.[4] As part of this shift towards thinking more carefully about places of science, scholars have also begun to consider how science is performed in particular locations. Central to this has been the work of Bernard Lightman, whose detailed accounts of popular scientific sites in London have been profoundly important to the field as a whole.[5] Iwan Morus and Dairmid Finnegan have also looked closely at science in performance, the latter in the lecture theatre and the former in public demonstrations.[6] Indeed, Morus's reflections on what it means to think of science as an activity or practice has provided an ever finer grained understanding of science spaces. For the most part, though, this work has looked in one direction; from the position of the scientist or scientific location outwards. Important questions have been asked from taking this standpoint, but it does mean that little attention has been paid to how scientific spaces were looked at, and represented, from the outside, from the perspective of the urban subject. For example, while Dierig, Lachmund, and Mendelsohn argue that one of the most important criteria for understanding science's place in urban space is to recognise that 'a city is also constituted by its representations', their ensuing discussion focuses on how scientists represented the city via maps, charts, and other regulatory mechanisms.[7] What they do not offer is any consideration of how the representations of the city might be accrued from representations of (not by) scientific sites, scientific activity, or scientists.

The central questions for this chapter are therefore how did science contribute to the vision of the city depicted by such guidebooks, and at the same time, in what ways did guidebooks mediate science in order to achieve this vision? These are new questions to ask about science in the Victorian period. While there has been extensive scholarship on the relationships between science and travel, and by extension travel writing, as well as on the relations between science and place, there has been almost none on the role of science in travel guidebooks. For example, Ralph O'Connor's rich and compendious work on popular geology and its representations includes a reference to only a single guidebook, and that in order to focus on its geologist author rather than the text. Aileen Fyfe's work on geology in travel guidebooks to the early decades of the nineteenth century stands alone in offering some analysis of the content of travel guides. Her interest, however, is in the ways that audiences engaged with geology, rather than in the guidebooks' own representations of science and what that might mean for science's place in nineteenth-century cultural formation.[8] We know, therefore, very little about the impact that guidebooks made on the way that science was received across Victorian Britain, and especially little of the major cities, where scientific activity was at its greatest intensity. Learning more about science in guidebooks to London will tell us how science was used in the creation of new forms of urban identity, both local and national. Moreover, we will come to understand the role science played in producing British urban culture, not only in terms of knowledge but as part of a wider self-fashioning of the British city and its inhabitants. We will also have the opportunity to uncover how travel guides constructed science. We will be able to consider how guidebooks to London staged science in textual formats, and how guidebook writers asked their readers to imagine that science. This work has significance, therefore, in expanding the canon of literature and science texts as well as in offering new territory for analysis in science studies more broadly.

Performance studies provides a useful way of reconstituting the direction of scholarly attention to science in the city. Advocates of performance studies, such as Richard Schechner, and the sociological perspectives that are a key influence upon the field, as reflected in the work of Erving Goffman, see the relationships between space and the subject (be that the human subject or another non-human actor) as a network of interrelated representations.[9] Goffman describes those relationships as assemblages of sign systems that can provide new meaning through 'dramatic realization'.[10] Performance studies is therefore more accepting of the role that the city's spaces, those who perform in them, and those who view them can play in co-creating meaning. For performance studies scholars, then, meaning might emerge from outside as well as inside particular spaces.

Travel guides are not performances, but they do work through a dynamic not dissimilar to that identified by performance studies. Travel guides represent science socio-spatially. They offer ways of viewing scientific spaces that are of course also constructions of the meaning of those spaces. But they also invite the reader to act out their role as tourist and to create meaning through performing for themselves. Moreover, travel guides allow their representations of science to be a part of a wider representation of the city itself (since all parts of the city go to constructing its identity, just as all aspects of performances make meaning). To study scientific representations in travel guides is to take a new perspective on 'the ways in which science moves within a city's interpretative communities', as Withers and Livingstone astutely recognise.[11] It is also to understand, in part, how science shifts its socio-spatial position as knowledge-maker and becomes urban culture.

THE PLACE OF SCIENCE IN TRAVEL GUIDEBOOKS

Science has always been closely associated with the production of travel narratives, and such narratives have often been regarded as literary products, too. The Romantic scholars, Tim Fulford, Debbie Lee, and Peter Kitson, for example, have written extensively about the relationships between science and exploration in late eighteenth-century travel narratives, highlighting the influence of James Cook's narrative, *A Voyage Towards the South Pole* (1777) and Mungo Park's *Travels to the Interior Parts of Africa* (1799) on scientific travel writing.[12] At the same time, and continuing into the nineteenth century, travel writing with science at its core had extended to descriptive tours of new industrial sites as well as to agricultural locations employing new industrial methods. As Benjamin Colbert and Judith Adler have both argued, travel guides like Thomas Young's *A Six Month Tour through the North of England* (1770) emerged as a new sub-genre that performed the act of travelling in order to comment on the social and political repercussions of scientific advancements.[13] Perhaps the best example of the literariness of travel writing about science is Charles Darwin's published narrative of his voyage on HMS *Beagle*. *The Voyage of the Beagle*, published in 1839 and explicitly a commentary on Darwin's scientific research, is also, as George Levine has shown, a self-conscious work of romanticism influenced by the British poets Darwin admired.[14] It is in this context that modern travel guides to locations in Britain first appeared: John Murray's first guide to London, for example, was published in 1849.

In their production and use travel guides were also imbricated with science and scientists. Murray's guides drew extensively on the expertise of the scientists whom Murray also published and which included writers of travel narratives like the archaeologist Henry Layard and later in the century the eugenicist Francis Galton. Guides such as *Bradshaw's Illustrated Handbook to London* (1862) came into existence specifically because of new technologies: George Bradshaw's expertise in travel publishing grew from his publication of railway timetables and his own involvement with civil engineering where accurate knowledge of civic space was essential to the engineer's enterprise.[15] Industrial events were also inspiration for the production of new travel guides. The Great Exhibition—or more accurately the Crystal Palace itself—generated numerous new travel guides to London, and an almost equal number of new editions of already existing guides. On smaller scales other national industrial exhibitions, such as the international exhibition in Edinburgh in 1888, saw a similar jump in the number of travel guides produced. Travel guides were also used by scientists to support their scientific work at home and abroad. The guides produced by Thomas Cook, for example, were central to the organisation of the archaeological surveys conducted by W.M. Flinders Petrie in Egypt in the 1880s.[16]

Travel guides were also written by scientists and increasingly by scientific institutions as well. Bradshaw's London guide may have been originally written by the dramatist Edward Blanchard, but later editions were collaborations with the agricultural analyst H. Kains Jackson, who conducted rural experiments for the Royal Agricultural Society while also writing reference works on Britain's agricultural territories. The scientific officers of the British Association for the Advancement of Science also caught the travel writing bug: not content with visiting Bristol for their annual meeting in 1875, the British Association also produced a travel guide to the city, *Bristol and Its Environs* (a title replicating many other existing guides to Britain's urban centres).[17] This guide may have been produced because of the Association's visit but it took on the title, format, and content of the modern travel guide and covered the major Bristol monuments as well as its scientific heritage. On a more modest scale other scientific institutions produced miniature guidebooks for public visitors (most likely in response to visitors being directed to them by the mainstream travel guides in the first place). George Heriot's Hospital School in Edinburgh, for example, copied the generic pattern of the larger travel guide in producing a booklet for visitors of its own history, architecture, and current activities which could be purchased by tourists acquiring a ticket to view the interior.[18]

Advertisers clearly recognised that travel guides appealed to readers who had an interest in science, as well as to tourists who wished to take advantage of medical science's pharmacological safety net whilst on holiday. Across a range of travel guides advertisements for medicinal products are plentiful, and almost always stress the necessity of good health whilst undertaking tourist activities away from home (and presumably from access to the family doctor). These medical adverts are joined by a further significant number for scientific instruments and technological objects. The well-known London instrument maker, Negretti and Zambra, to give one example, took a full-page advertisement in *Cook's Handbook for*

London for 1878, in order to promote the firm's binoculars, barometers, and other optical instruments that might be seen as 'tourist's scientific requisites'.[19] Additionally, several travel guides, and especially those with extensive publishing interests in other genres, also advertised science's textual culture through the promotion of scientific works that would add to and enhance a tourist's knowledge of the scientific sites listed in the guides. *Black's Guide to Edinburgh* included in its list of works published by fellow Scottish publisher Blackwood's David Page's *Introductory Text Book of Geology* (first published in 1854 but extensively reprinted), which would supplement Black's own recommendations for geological tourism at Edinburgh's well-known igneous rocks called Salisbury Crags, first examined and defined by James Hutton in the 1780s.[20]

Science was present at every stage of travel guide production and use: from the conception of a guidebook through its months of research, into its highly collaborative writing practices, to final choices about content, and in taking account of a perceived readership.[21] One clear reason for this is science's extensive role in the formation of British culture. Travel guides are both repositories of culture (at least of a moment of culture) and producers of culture. Those focused on urban spaces therefore reflect the city and make the city simultaneously. In doing so they inevitably have to capture the roles that science has played in the intellectual and material life of the city throughout its history and particularly in the present. In addition to this, science as a set of disciplines dedicated to new knowledge has something in common with the travel guide and the tourist reader. Both science and the tourist privilege discovery, or at least the idea of discovery as an epistemological principle of the activity they choose to undertake. Science offers the travel guide writer the opportunity to reflect the tourist's own desire for the new in the places, spaces, and actors that they represent. In this way science becomes an alternative route towards the kind of authenticity that critics see as an essential part of the tourist or traveller experience. Alex Milsom summarises this critical perspective well in noting that scholars read authentic tourist experiences as those that offer 'a refuge from the vastly changing world' and are therefore often equated with ancient sites that seem, in their very age, to connect to a truer, more stable, past.[22] Contradictorily, science in travel guides provide an authenticity of experience detached from the past and focused on an engagement with the excitement of contemporaneous modernity. Indeed, as we shall see, in travel guides for Victorian London science performs a role that guides the tourist not just away from the past but into the future.

PRESENTING SCIENCE IN VICTORIAN LONDON

Travel guides to London, across the second half of the nineteenth century, offer a vision of the city through the sciences that is, if not entirely uniform, quite remarkably consistent. In their presentation of science, travel guides to London are also quite distinct from travel guides to other cities across Britain. In this section I shall discuss five of the most important travel guides to Victorian London: *Bradshaw's Illustrated Handbook to London* (1862), *Cook's Handbook for London* (1878), Baedeker's *London and its Environs* (1881), Routledge's *London and its Environs* (1888), and *Black's Guide to London and its Environs* (1891). These guides cover four decades from the second half of the nineteenth century and also take in both British and European perspectives. Three of the guides are produced by British writers and publishers, Baedeker's guide emerges from a German context while Routledge's is translated directly from French and gives a Parisian view of the city. All of the guides focus on a similar, although not identical, range of scientific sites within London. These include places of active scientific work such as the Greenwich Observatory, the Royal Institution or the Woolwich Arsenal, sites like the South Kensington Museum, the Royal College of Surgeons or the Crystal Palace where scientific and technological knowledge is organised and displayed, and places where science merges with public entertainment, as at the Zoological Gardens in Regent's Park or Kew Gardens at Richmond. Finally, each guide focuses too on examples of applied science: on bridges built with particular engineering skill or the extent of the telegraph system.

What becomes readily apparent in reading the travel guides' itineraries for London walks, studying their historical introductions, or browsing their lists of key tourist locations (activities which the guides encourage and which is likely to be the way that contemporary tourists first consumed the guides they bought) is that science is ubiquitous. *Cook's Handbook* is organised as an alphabetical gazetteer of London sights, with categories such as transport, public buildings, theatres, and museums. It lists half a dozen scientific museums, five public buildings dedicated to scientific or medical pursuits, two pages of hospitals, and four scientific places of amusement, from a total of eight across the city.[23] *Black's Guide*, published in the final decade of the nineteenth century, chooses not only to list numerous scientific sites for its readers but also to draw attention to how tourists should think of them. 'The importance of the SCIENTIFIC AND ART COLLECTIONS of London', notes the guide's Preface, 'requires that special attention should be directed to them'.[24] *Black's* demand that tourists should mobilise in order to attend specifically to the sciences, rather than encounter them within the wider remit of a walking tour, gives to science the status of the required destination, placing it atop the hierarchy of tourist sites of significance.

While *Black's* guide and others, either explicitly or implicitly, indicate that to come to any knowledge of London one must understand its science, travel guides also employ science in a more performative role. Both Routledge's *London* and *Bradshaw's Handbook* cite a scientist in service of a claim about the city's global significance. For Pierre Villars and Henry Frith, writing and translating the Routledge guide, this came at the conclusion to the various walking tours of London: 'In the foregoing pages we have indicated in broad lines the curious, complex, and ever-changing features of the gigantic town which Herschel considered the centre of the universe'.[25] Blanchard, writing *Bradshaw's Handbook*, preferred to begin with Herschel's quip: 'Within the last fifty years, London has more than doubled in extent; and, even as we write, is rapidly increasing in every direction. It was happily observed by Herschel, that London occupies nearly the centre of the terrestrial hemisphere—a fact not a little interesting to Englishmen'.[26] Both guides wish to exploit the Swiftian humour inherent in a leading British astronomer placing London at the centre of the universe. Yet Herschel is playing another role than that of the satirist. His view of London may raise a smile from the tourist reader, but it also carries the authority of his astronomical knowledge. The very fact that it is Herschel who has placed London so centrally in the universe makes the comment less comic and much more likely to be true: the tourist reader may well bow to Herschel's greater understanding of the systems of the heavens. If so, London is transfigured by Herschel's claim; no longer only the imperial city it is now the most significant of sites in a much larger cosmological vision of humanity. Indeed, Villars, Frith, and Blanchard continue to hint at astronomical influence: they describe London as 'ever-changing' and 'rapidly increasing', just as astronomers were focusing attention on the expanding universe. By using a vocabulary that is common to both urban experience and astronomical research they position London as the universal city.

Travel guidebooks to London use other aspects of contemporary science to continue this theme of representing the city as a place that is both massive and monumental. David Gilbert, in his analysis of how travel guides promote London's imperial character, notes that many guides 'betrayed anxiety that London was insufficiently spectacular or appropriately "imperial" in its urban form'.[27] In particular, Gilbert notes that London was seen as lacking the kinds of historic monuments, and the vistas by which to view them, enjoyed by other imperial cities. This was in contrast, particularly, to Paris, where tourist guides, as Margaret Cohen has argued, offered a decidedly 'monumental' set of itineraries for their readers.[28] Routledge's *London*, written for a French audience and always with one eye on comparisons between London and Paris, writes directly about this issue. London, Villars writes, reveals an 'almost entire absence of ancient monuments' but this is explained by the city's decision to 'sacrifice all the ancient monuments which [...] impede the development of their metropolis'.[29] Following Villars's itineraries as he navigates the tourist through London's monument-less cityscapes, it becomes apparent that the sacrificed ancient monuments have been replaced by more recent structures. Many of these host scientific activities or are themselves scientific spaces. The Royal Institution, which is described specifically as 'recent' is one such building, and Villars draws particular attention to one aspect of its popular scientific work—its appeal to 'persons connected with industry and trade'.[30] Indeed, throughout the Routledge guide it is science's support for the engines of imperialism (industrial might and global trade) that is stressed, enabling scientific sites to perform as monuments to empire in lieu of others of a more historic or civic function. The engineering found in London's docklands is 'a newly discovered world', the new South Kensington museums (only partially completed by 1888) are described as celebrating 'industrial art' and the numerous new scientific institutions of which London boasts are presented as having 'uprisen from very small beginnings' as a result of 'private enterprise'.[31] All of these scientific sites try to turn science into commerce, or at least to apply it to commerce in order to give a coherent narrative of empire built upon monumental knowledge.

Routledge's *London* is not alone in constructing science as monumental in order to present London as a city associated with great monuments. Both historic and contemporary scientific sites are chosen to construct a vision of London as a city of unique significance, a city where the tourist will find, as *Bradshaw's Handbook*, drawing on Shakespeare's *The Tempest*, said, 'wonders begin to accumulate around him'.[32] For Blanchard, writing *Bradshaw's Handbook*, the Crystal Palace was the most important new monument to science. The tourist will find the Crystal Palace 'surpassing all conception' he claims, and come to see it as an incredible work 'of human ingenuity'.[33] Even Baedeker's *London*, largely written by the art historian Jean Paul Richter and unsurprisingly leaning towards London's galleries, sees it as nothing less than 'magnificent'.[34] Although the Crystal Palace is clearly the most eminent of all scientific sites, and across almost every guidebook of this period, other historic buildings are also brought to the attention of the tourist seeking evidence of London's monumental culture. Somerset House is often drawn in to play this role: guidebooks note its architectural beauty and its royal heritage before noting the continuation of its pre-eminence in its role as host to the savants of the Royal Society. Even *Bradshaw's Descriptive Railway Handbook for Great Britain*, summarising Blanchard's London guide in very limited space, takes the time to describe Somerset House as of 'striking magnificence' and 'the finest object of the kind in London'.[35] After the Royal Society moves from Somerset to Burlington House in 1874 subsequent guides give very little time to depicting Somerset House. *Black's Guide*, the latest of those considered here, notes only its name while giving more than a column to describing the scientific objects that can now be viewed at Burlington House.[36] Science, it appeared, was becoming an essential feature of the contemporary city monument.

Alongside the new monumentalising of London, science was also called upon by guidebook writers to signify the city's scale. London was, according to *Black's Guide*, 'the largest city in the world',[37] or, as *Bradshaw's Handbook* more suggestively told its readers, nothing less than an urban 'Leviathan'.[38] While these claims were made in the general introductions to London, which were designed to give tourists a sense of the importance of the city and of the excitement of visiting it and were therefore traditionally extravagant, there was still a necessity to provide some sense of the truth of the claim in the subsequent itineraries. Various scientific sites were brought to bear to deliver practical examples of the city's scale, not only the size of individual sites or of buildings and parks, for example, but also the more elusive sense that London was large enough to offer everything that a tourist might desire. The Crystal Palace was commonly employed as a key indicator of both size and variety: although such an example would already have been obvious to most visitors. Other scientific locations, such as the Royal Institution, also gave guidebooks the opportunity to reveal the hidden scale of London. *Bradshaw's Handbook* describes the Royal Institution first as 'famous for its weekly lectures on chemical science' before describing its 'handsome façade' on Albemarle Street. The real significance of the Royal Institution, though, was its laboratory, which was 'fitted up on a scale of magnitude and completeness never before attempted in this country'.[39] Science is here employed to perform a novelty of scale for the newly enormous city and additionally to stand as a marker for the enormity of ambition of the nation as a whole.

One site that is continually produced to give a sense of the scale of London, and which simultaneously gives a sense of the scale of its applied science, is the military research site at the Woolwich Arsenal. *Cook's Handbook*, organised as a gazetteer rather than an itinerary guidebook, calls the Arsenal simply the 'largest in the world' (a familiar guidebook refrain about London itself) but most guides give greater detail.[40] Routledge's *London* follows *Cook's Handbook* in its initial assessment of the Arsenal as 'an immense establishment' but particularly stresses the hugeness of the laboratories, one of which contains '500 lathes, and 1200 men can work there comfortably'.[41] Baedeker's *London*, also written with a non-British audience in mind, is even more impressed with the Arsenal's scale as 'one of the most imposing establishments in existence for the manufacture of materials of war'. In particular, Baedeker's *London* is awestruck by the magazines, which 'extend along the Thames for nearly a mile'.[42] The Arsenal, for both these guidebooks, is also very much a national institution. Both guidebooks stress the difficulty of accessing the site as a foreign tourist. To do so requires 'a special permit', Routledge's *London* advises, and even then it is only obtained with 'some little difficulty'.[43] While the guidebooks clearly play on the excitement offered by the hint of military espionage enjoined to restricted access, they also clearly denote the Arsenal as a British space that is exclusive to the nation's citizens. The science of the Woolwich Arsenal is British science: it connotes the national characteristics of industry and indefatigability as well as inventiveness. The Arsenal also occupies a key place in the iconography of London as a city of immensity and of man-made wonder.

It is a British guidebook that drives home this performance of national identity most assiduously. Blanchard, in *Bradshaw's Handbook*, spends considerable space (more than a full page) on the Arsenal and its description is worth giving in detail:

> The 'Royal Arsenal' will be observed but a short distance off, composed of several buildings, wherein the manufacture of implements of warfare is carried on upon the most extensive scale. On entering the gateway the visitor will see the 'Foundery' before him, provided with every thing necessary for

casting the largest pieces of ordnance, for which, as in the other branches of manufacture, steam power has been lately applied. Connected with the 'Pattern Room,' adjoining, will be noticed several of the illuminations and devices used in St. James's Park to commemorate the peace of 1814. The 'Laboratory' exhibits a busy scene, for here are made the cartridges, rockets, fireworks, and the other chemical contrivances for warfare, which, though full of 'sound and fury,' are far from being considered amongst the enemy as 'signifying nothing.' To the north are the storehouses, where are comprised outfittings for 15,000 cavalry horses, and accoutrements for service. The area of the Arsenal includes no less than 24,000 pieces of ordnance, and 3,000,000 of cannon-ball piled up in huge pyramids. The 'Repository' and 'Rotunda' are on the margin of the common, to the south of the town, and contain models of the most celebrated fortifications in Europe, with curiosities innumerable.[44]

As with other guidebook representations of scale, *Bradshaw's Handbook* here gives numerous numbers in an effort to impose arithmetical astonishment upon the tourist reader. It also suggests the size of the Arsenal and the wonder associated with that, particularly in describing the cannon-ball as pyramids and the limitless curiosities of the models. But it is the manner in which the description of applied chemistry is tied in to events of national significance and to national texts (Shakespeare's *Macbeth*) that is striking. What the guidebook writer draws on here are two performances. The first is a national pageant organised to mark the Peace Treaty signed by Britain at the conclusion of the Napoleonic Wars. The second is the quotation from Shakespeare's *Macbeth*, which in its original context is spoken at the end of Macbeth's soliloquy on the similarity between life and a staged performance.[45] Of course Blanchard's own career as a dramatist may well have led him to choose a well-known dramatic speech to illuminate the Arsenal's scientific achievements. Nevertheless, he does, wittingly or not, imbricate London with the nation and science with the performance of national identity. His description of the Arsenal suggests to the visiting tourist that Britain's chemical knowledge contributes to the making of the nation and its capital city but might also perform the very acts of nation-building that contribute to a sense of London as both part of and representative for a broader Britain that reaches beyond the immediate cityscape.

Representing London as a city that brings things together within the urban space, and might even make them complete, is also achieved by focusing on the city's sciences. Across every guide to London, from the

1850s to the end of the century, there is discussion of the city's extensive scientific collections. From the British Museum's collections of archaeological remains through the Royal College of Surgeon's anatomical collection and on to the South Kensington natural history collections objects gathered together for the purposes of scientific study are always pointed out to tourists as essential for visiting. These collections very often double as examples of the scale of London and its sciences. Blanchard, for example, stressed the collection and scale of the British Museum's books: 'assuming the shelves to be filled with books, of paper of average thickness, the leaves placed edge to edge would extend about 25,000 miles, or more than three times the diameter of the globe'.[46] In the main, though, collections were used to distinguish London as an urban space that unified knowledge and brought together objects more commonly disparate from one another.

Black's Guide saw the Zoological Gardens in Regent's Park as the exemplary model of London's enthusiasm for collecting. The Zoo offered the tourist 'the most complete collection of live specimens in the world'.[47] *Bradshaw's Handbook of Great Britain* agreed. The Zoological Gardens displayed 'specimens of rare, curious, and beautiful animals' which had been 'collected from every corner of the globe [...] with great care and order'. The result, achieved by 'that spirit of association which has achieved so much for England', was that 'a walk through this garden is in a measure like a rapid journey over the world'.[48] For these and other guidebooks offering a similar description of the Zoological Gardens and other natural history sites, the Zoo captured something of the essential nature of London and of British character.[49] At the Zoo, London became a city that could reach across the globe, but which also depended on home-grown initiative and collaboration. Both *Black's Guide* and *Bradshaw's Handbook* took their tourist readers directly from the Zoological Gardens out onto Primrose Hill for a panoramic view of the city. It seems fitting to do so: both the Zoo and the view depict London as singular and whole, a complete city that identifies with completion and whose scientific activities reflect and define the desire for unity.[50]

The sense of scale and the designs upon systematic completeness that the travel guidebooks offer to the visiting tourist hint at London as a city with a utopian vision of its own future. Lynda Nead has argued that from around the middle of the century through the later decades 'London became part of a highly concentrated discourse on the modern'.[51] One aspect of that modernity—and the aspect which the sciences most keenly express—is the possibility of imagining the future. It is perhaps not surprising that travel guidebooks employ science to consider the future, but it is surprising that travel guidebooks give a sense of the future at all. It is much more common to view travel guides as fixed upon the past. Their production and generic features—publication was often months or years after the writer had visited a location and they tended to rely on earlier guidebooks for guidance on the key sites to include—meant travel guides were always writing about historic features. This was especially true of Victorian cities, which were changing reasonably quickly, and certainly more quickly than did new editions of the most popular guides.[52] Despite this, guidebooks to London did point towards a future, even if that was only in the most general terms as a city of 'novelty, and change'.[53]

Some guidebooks attempted to say more about the London of the future, and when they did they offered a utopian version of a city made into paradise by employing scientific knowledge to its fullest effect. Baedeker's *London* paused while showing the tourist around the South Kensington museums to reflect that 'among its many professors, directors, and examiners are numbered many of the chief English *savants*' and that this was driving London towards the advancement of science and art across Britain: 'the tangible results of its teaching and influence are seen in the progress of taste and knowledge in the fine arts and natural science throughout the kingdom'.[54] Routledge's *London* saw the results of scientific advancement in terms of social progress rather than refinement. London was likely to become, Villars argues in his conclusion to the Routledge guidebook, 'the most beautiful city in Europe, as it is already the largest and most healthy'. That health had emerged, of course, because of technological improvements, new public health knowledge, and the continual progress of civil engineering. In a final comment that rather brilliantly undermines early twenty-first century perspectives on the nineteenth-century city, Villars invites British residents of London to 'compare their formerly dark and smoky city with the London of the Victorian era'.[55]

Bradshaw's Handbook, written 25 years earlier, provides a very good example of what, for Villars, had made such a difference to the London of the 1880s. Blanchard writes of the New Westminster Bridge, which had recently (in 1860) been demolished and rebuilt, employing cutting-edge engineering technologies and new chemical innovations. The bridge is particularly improved, Blanchard notes, by 'being very nearly level with the approaches on both sides' which 'affords great relief' to both pedestrians and other traffic. The bridge will, Blanchard enthuses, 'deserve a critical visit' by the tourist.[56] Especially of note is the new lighting:

> The stranger should observe that the completed portion of the new bridge is lighted by the lime light, lately introduced. Ten lights, about one-third of the number of the old gaslights, present a most brilliant appearance. It is most interesting to know that in the lime light Newton's assertions are fully corroborated. The oxy-hydrogen flame burns the constituents of water, and water is the only product of such combustion. The chief feature and improvement in these lamps is the adaptation of lime as the reflecting surface on which the jet of flame from coal gas plays, and which becomes intensified to an extraordinary degree, and making the old gas burners, in close proximity, appear dull, as though they were burning in the day time.[57]

Blanchard sees in Westminster Bridge the London of the future appearing from the innovations of science. As Villars was to do in his Routledge guide, Blanchard uses darkness as a metaphor for the past, while brightness and illumination are markers of the future. Blanchard's description of the New Westminster Bridge is interesting in the context of Lynda Nead's reading of the semiotics of gas lighting in this period. Nead argues that gas signified modernity and metropolitan culture on the one hand and 'industrial production and urban blight' on the other.[58] Blanchard, focusing on scientific advancements of several kinds, rejects that separation by seeing large-scale engineering positively, as necessary in the shift towards the future.

Conclusion: Science Performing London

One caveat to this reading of Blanchard's socio-spatial depiction of Westminster Bridge is his decision to employ Newton as a symbol of Victorian science. What possible role could an eighteenth-century natural philosopher play in depicting nineteenth-century science as cutting-edge futurism, even if that philosopher was the very definition of the kind of genius that the British would wish to claim in the self-fashioning of their national identity?[59] Walter Benjamin would not see this as incongruous. He would argue that this collision between present and past is the very essence of the urban experience. In his influential analysis of nineteenth-century Paris, Benjamin sees a desire amongst the new capitalist class to seek out 'the primal past' whenever they wish to dream of the new epoch they might create, so that these two things—the immediate present and

the distant past—appear always 'coupled'.[60] The spectre of the past haunt-
ing the present, as Nead portrays Benjamin's analysis, has become a standard way of understanding the mobility of the cityscape and particularly
of its phenomenological iteration in the personal urban experience.[61]
Benjamin's view has been reinforced by theorists of urban performance.
De Certeau, in his reading of everyday life in the city, sees commonplace
urban tours as spatialising practices that exist as 'not a "geographical map"
but [a] "history book"'.[62] Here, too, urban places are arrested by the past
rather than by any desire to locate them spatially.

This appears to be how Blanchard gives his own cultural co-ordinates
for the new Bridge. The Bridge is of course *Westminster* Bridge, and it is
encountered in an itinerary that organises the tourist in space. Yet in viewing the Bridge its *Westminster-ness* fades into the background in favour
of Newton's place in its ongoing history. It becomes Newton's bridge:
its location moving away from spatial containment into the mythology
of history. Blanchard exacerbates this sense of the importance of history
over location by quoting in full Wordsworth's sonnet on the London sun-
rise, a sonnet composed, as its title tells us, upon Westminster Bridge.[63]
However, Blanchard quotes it several itineraries later, while the tourist is
placed on Waterloo Bridge. Wordsworth, like Newton, is an essential part
of the history of the city, but clearly not of its spatial organisation.

Yet, the representation of Newton guards against tourists viewing him
and his scientific achievements as remnants of science's past still casting a
spell over the Victorian present. Newton, whose 'assertions' of the properties of lime light are only 'fully corroborated' in the 1860s, is cast as the seer
(indeed seeing clearly is a vital undercurrent to the reading of Westminster
Bridge); he is a prophet of the future, whose scientific hypotheses illuminate a future period able to understand and implement them. Newton
looks forward, then, to an altogether brighter future, a future inhabited
by Blanchard and by the tourist reader standing upon Westminster Bridge
consulting their Bradshaw. This is the performance that Newton offers to
the travel guidebook reader, and it is one that other travel guidebooks
replicate. In his slightly earlier guide to London, *The Town* (1848), the
critic and essayist Leigh Hunt spoke of Newton in the context of the Royal
Society, then at Somerset House. Hunt begins by noting Newton's role in
bringing to an end the public ridicule of the Royal Society's experiments.
When Newton became its President, Hunt notes for his tourist readers,
'jesting ceased'. This reminds Hunt once again of the present Royal Society,
which his itinerary has taken the tourist past: 'It is pleasant to think, while

passing Somerset House, in the midst of the noise of a great thoroughfare, that philosophical speculation is, perhaps, going on within these graceful walls'. It is exciting, he concludes, for the tourist to imagine themselves 'in the midst of [all] sorts of new things'.[64] The Newton of Hunt's travel guide is another seer of the future, recognising in the Royal Society an organisation of future potential that will at some point discover 'new things' and stepping in to enable that to happen. Like *Bradshaw's Handbook*, Hunt's *The Town* suggests that this future might be both the tourist readers' present moment and a further future where scientific work will still be 'going on' in the heart of London. Newton, then, is no ghostly presence from the past but instead an icon of futurity, a symbol of science's utopian potential and simultaneously of London's far-sightedness and inventiveness.

To dream of the future does not mean that each age must look to its past, as Benjamin claims in his reading of Michelet's famous aphorism.[65] Indeed the example of science considered through a range of travel guidebooks to London shows how the past could be harnessed to support a vision of the future rather than detain it. While we might think of the second half of the nineteenth century as a time when scientific sites and the scientists who inhabited them were attempting to present themselves anew—as offering sober and professional expertise—it is clear that alternative performances of science were increasingly available in places and texts over which scientific communities had little control.[66] Looking closely at the depictions of science in travel guidebooks is also a reminder of how mutually constitutive science and location are. While science in London is clearly defined for tourists as impressive in scale, in completeness, and vital in creating a utopian future it is also the case that London, too, gains in definition. London becomes a city of the future, its urban landscape one of scientific and technological marvels designed to enhance human life. It is also possibly in these travel guides that scholars too can look a little into the future and see, from one perspective at least, the first signs of science's future position in civic culture: to the late twentieth and early twenty first century where science's pre-eminence as the only field of knowledge able to tackle international challenges, to find for humanity a better future, is often lauded and rarely challenged. As the arguments presented here have shown, literary and artistic presentations of science have huge value in promoting not only particular representations of science, but the idea of science as both innovatively contemporary and constitutive of a future modernity that will maintain British intellectual pre-eminence. It should be of considerable importance to literature and science scholarship that it is humanist practice—the writing of guidebooks and dramatising of cityscapes—that underpins this achievement.

NOTES

I would like to express my thanks to Simon Avery, David Cunningham, Alex Warwick and Ann Heilmann for reading and commenting on early versions of this chapter.

1. Erving Goffman, *The Presentation of Self in Everyday Life* [1959] (London: Penguin, 1990).
2. Sven Dierig, Jens Lachmund and J. Andrew Mendelsohn, 'Introduction: Toward an Urban History of Science', *Osiris*, 18 (2003), 1–19 (p. 2).
3. Charles W.J. Withers and David N. Livingstone, 'Thinking Geographically about Nineteenth-Century Science', in *Geographies of Nineteenth-Century Science* ed. David N. Livingstone and Charles W.J. Withers (Chicago: University of Chicago Press, 2011), 1–19 (p. 2).
4. Henri Lefebvre, *The Production of Space* (Oxford: Blackwell, 1991); Michel De Certeau, *The Practice of Everyday Life*, trans. Steven Rendall (Berkeley: University of California Press, 1984).
5. Bernard Lightman, *Victorian Popularizers of Science: Designing Nature for New Audiences* (Chicago: University of Chicago Press, 2007). See also Lightman's essay in *Geographies of Nineteenth-Century Science* ed. Livingstone and Withers, 25–50.
6. Iwan Rhys Morus, 'Placing Performance', *Isis*, 101 (2010), 775–778. Morus's article introduces a Focus Section entitled 'Performing Science' which he conceived and edited. Dairmid A. Finnegan, 'The Spatial Turn: Geographical Approaches in the History of Science', *Journal of the History of Biology*, 41 (2008), 369–388.
7. Dierig et al, p. 10.
8. Aileen Fyfe, 'Natural History and the Victorian Tourist: From Landscapes to Rock-Pools' in *Geographies of Nineteenth-Century Science* ed. Livingstone and Withers, 371–398. In constructing an argument about what audiences received from geological writing in travel guidebooks, Fyfe's work also takes its place in the scholarship that looks in one direction—from science and into culture. See also Ralph O'Connor, *The Earth on Show: Fossils and the Poetics of Popular Science, 1802–1856* (Chicago: University of Chicago Press, 2007).
9. Richard Schechner, *Performance Studies: An Introduction* (New York: Routledge, 2002); Goffman, *Presentation*. See also *Performance Studies* ed. Erin Striff (Basingstoke: Palgrave Macmillan, 2003) and *The Performance Studies Reader* ed. Henry Bial (New York: Routledge, 2004).

10. Goffman, *Presentation*, p. 40.
11. Withers and Livingstone, 'Thinking Geographically', p. 6.
12. Tim Fulford, Debbie Lee and Peter Kitson, *Literature, Science and Exploration in the Romantic Era* (Cambridge: Cambridge University Press, 2004).
13. Benjamin Colbert, 'Aesthetics of Enclosure: Agricultural Tourism and the Place of the Picturesque', *European Romantic Review*, 13.1 (2002), 23–34; Judith Adler, 'Travel as Performed Art', *American Journal of Sociology*, 94.6 (1989), 1366–1391.
14. George Levine, *Darwin the Writer* (Oxford: Oxford University Press, 2011), pp. 5–6.
15. Edward Blanchard, *Bradshaw's Illustrated Handbook to London and its Environs* (London: W.J. Adams, 1862).
    Bradshaw's Illustrated Handbook to London and its Environs (London: W.J. Adams, 1862)
16. Martin Willis, *Vision, Science and Literature, 1870–1920: Ocular Horizons* (London: Pickering and Chatto, 2011), pp. 115–141.
17. *Bristol and its Environs: Historical Descriptive and Scientific* (London: Houlston and Sons, 1875).
18. *Guide to George Heriot's Hospital, Edinburgh* (Edinburgh: Bell and Bradfute, 1872).
19. 'Negretti and Zambra, Tourist's Scientific Requisites', *Cook's Handbook for London* (London: Thomas Cook, 1878), p. 108.
20. *Black's Guide to Edinburgh and Environs* (Edinburgh: Adam and Charles Black, 1871).
21. Archival materials relating to John Murray's travel guides from the 1830s throughout the nineteenth century—letters, annotations to proofs, internal memoranda, and financial records—indicate an extensive and wide-ranging set of collaborations between the scientists who were contracted to write for Murray on a number of different projects and the travel guides written either by Murray and his associates or by commissioned authors. See John Murray Archive, National Library of Scotland.
22. Alex Milsom, '19th-century Travel and the 21st-century Scholar', *Literature Compass*, 10.9 (2013), 725–733 (p. 728).
23. *Cook's Handbook*, pp. 46–79.
24. *Black's Guide to London and its Environs* (London: Adam and Charles Black, 1891), pp. v–vi.
25. Pierre Villars and Henry Frith, *London and its Environs* (London: George Routledge and Sons, 1888), p. 191.
    London and its Environs (London: George Routledge and Sons, 1888).
26. Blanchard, *Bradshaw's Handbook*, p. 14.

27. David Gilbert, '"London in all its glory – or how to enjoy London": Guidebook Representations of Imperial London', *Journal of Historical Geography* 25.3 (1999): 279–297, p. 284.
28. Margaret Cohen, *Profane Illumination: Walter Benjamin and the Paris of Surrealist Revolution* (Berkeley: University of California Press, 1993), p. 77.
29. Villars and Frith, Routledge's *London*, p. 18.
30. Villars and Frith, Routledge's *London*, p. 122.
31. Villars and Frith, Routledge's *London*, pp. 49, 119, 120.
32. Blanchard, *Bradshaw's Handbook*, p. 17.
33. Blanchard, *Bradshaw's Handbook*, p. 215.
34. Jean Paul Richter, *London and its Environs* (Leipsig: Karl Baedeker, 1881), p. 291.
    London and its Environs (Leipsig: Karl Baedeker, 1881).
35. *Bradshaw's Descriptive Railway Handbook of Great Britain and Ireland* (London: W.J. Adams, 1863), p. 7.
36. *Black's Guide to London*, pp. 283, 316–317.
37. *Black's Guide to London*, p. 1.
38. Blanchard, *Bradshaw's Handbook*, p. 15.
39. Blanchard, *Bradshaw's Handbook*, pp. 112–113.
40. *Cook's Handbook*, p. 46.
41. Villars and Frith, Routledge's *London*, p. 204.
42. Richter, Baedeker's *London*, p. 289.
43. Villars and Frith, Routledge's *London*, p. 204.
44. Blanchard, *Bradshaw's Handbook*, p. 191.
45. Macbeth's speech goes as follows: 'Life's but a walking shadow, a poor player/That struts and frets his hour upon the stage/And then is heard no more: it is a tale/Told by an idiot, full of sound and fury,/Signifying nothing.' (Act 5, Scene 5).
46. Blanchard, *Bradshaw's Handbook*, p. 145.
47. *Black's Guide to London*, p. 94.
48. *Bradshaw's Handbook of Great Britain*, p. 10.
49. Kew Gardens is the other key London scientific site that achieves this. *Bradshaw's Handbook*, for example, describes it as so impressive and complete that 'a volume would be required to describe its attractions' (p. 174).
50. This is reinforced by the denigration of sites of science that are incomplete in some way. *Black's Guide* criticises the Royal Institution as an 'amateur' society primarily because its lectures are 'not usually of a systematic character.' (p. 319).
51. Lynda Nead, *Victorian Babylon: People, Streets and Images in Nineteenth-Century London* (New Haven: Yale University Press, 2000), p. 5.

52. David Webb, 'For Inns a Hint, for Routes a Chart: The Nineteenth-century London Guidebook', *London Journal*, 6.2 (1980), 207–214 (p. 210).

53. Blanchard, *Bradshaw's Handbook*, p. 18.

54. Richter Baedeker's *London*, p. 262.

55. Villars and Frith, Routledge's *London*, p. 192.

56. Blanchard, *Bradshaw's Handbook*, pp. 84–85.

57. Blanchard, *Bradshaw's Handbook*, p. 85.

58. Nead, p. 97.

59. Newton was born in 1643 and died in 1727, in London. He could be regarded as both a seventeenth-century and an eighteenth-century figure.

60. Walter Benjamin, 'Paris—Capital of the Nineteenth Century' in *The Arcades Project* (Cambridge: Harvard University Press, 2002): 77–88, p. 79.

61. Nead, *Victorian Babylon*, p. 6.

62. De Certeau, *Everyday Life*, p. 119.

63. William Wordsworth 'Composed upon Westminster Bridge, Sept. 3rd, 1803' (1806) is quoted in *Bradshaw's Handbook*, p. 169.

64. Leigh Hunt, *The Town* (London: Smith, Elder and Co, 1848), pp. 170–171.

65. Benjamin quotes from Michelet in *The Arcades Project*: 'every epoch dreams its successor' (p. 79).

66. Bernard Lightman, 'Refashioning the Spaces of London Science: Elite Epistemes in the Nineteenth Century' in *Geographies of Nineteenth-Century Science* ed. Livingstone and Withers, 25–50 (p. 36). Archival materials on the production of Murray's Handbooks reveal that scientists did contribute to travel guidebooks and it is therefore not possible to see the scientific and travel writing communities as separate. See John Murray Archive, National Library of Scotland, Ms. 42613, Ms. 42727–42732.

BIBLIOGRAPHY

Adler, Judith, 'Travel as Performed Art', *American Journal of Sociology*, 94.6 (1989), 1366–91.

Benjamin, Walter, *The Arcades Project* (Cambridge: Harvard University Press, 2002).

Bial, Henry, ed., *The Performance Studies Reader* (New York: Routledge, 2004).

Black's Guide to Edinburgh and Environs (Edinburgh: Adam and Charles Black, 1871).

Black's Guide to London and its Environs (London: Adam and Charles Black, 1891).

Bradshaw's Descriptive Railway Handbook of Great Britain and Ireland (London: W.J. Adams, 1863).

Bristol and its Environs: Historical Descriptive and Scientific (London: Houlston and Sons, 1875).

Cohen, Margaret, *Profane Illumination: Walter Benjamin and the Paris of Surrealist Revolution* (Berkeley: University of California Press, 1993).

Colbert, Benjamin, 'Aesthetics of Enclosure: Agricultural Tourism and the Place of the Picturesque', *European Romantic Review*, 13.1 (2002), 23–34.

Cook's Handbook for London (London: Thomas Cook, 1878).

De Certeau, Michel, *The Practice of Everyday Life*, trans. Steven Rendall (Berkeley: University of California Press, 1984).

Dierig, Sven, Jens Lachmund and J. Andrew Mendelsohn, 'Introduction: Toward an Urban History of Science', *Osiris*, 18 (2003), 1–19.

Finnegan, Dairmid A., 'The Spatial Turn: Geographical Approaches in the History of Science', *Journal of the History of Biology*, 41 (2008), 369–88.

Fulford, Tim, Debbie Lee and Peter Kitson, *Literature, Science and Exploration in the Romantic Era* (Cambridge: Cambridge University Press, 2004).

Fyfe, Aileen, 'Natural History and the Victorian Tourist: From Landscapes to Rock-Pools' in *Geographies of Nineteenth-Century Science*, eds David N. Livingstone and Charles W.J. Withers (Chicago: University of Chicago Press, 2011).

Gilbert, David, '"London in all its glory – or how to enjoy London": Guidebook Representations of Imperial London', *Journal of Historical Geography*, 25.3 (1999), 279–97.

Goffman, Erving, *The Presentation of the Self in Everyday Life* [1959] (London: Penguin, 1990).

Guide to George Heriot's Hospital, Edinburgh (Edinburgh: Bell and Bradfute, 1872).

Hunt, Leigh, *The Town* (London: Smith, Elder and Co, 1848).

John Murray Archive, National Library of Scotland, accessed via <http://www.nls.ac.uk>.

Kirshenblatt-Gimblett, Barbara, 'Performance Studies', in *The Performance Studies Reader*, ed. Henry Bial (New York: Routledge, 2004), pp. 43–53.

Lefebvre, Henri, *The Production of Space* (Oxford: Blackwell, 1991).

Levine, George, *Darwin the Writer* (Oxford: Oxford University Press, 2011).

Lightman, Bernard, *Victorian Popularizers of Science: Designing Nature for New Audiences* (Chicago: University of Chicago Press, 2007).

Lightman, Bernard, 'Refashioning the Spaces of London Science: Elite Epistemes in the Nineteenth Century', in *Geographies of Nineteenth-Century Science* eds Livingstone and Withers (Chicago: University of Chicago Press, 2011), pp. 25–50.

Milsom, Alex, '19th-century Travel and the 21st-century Scholar', *Literature Compass*, 10.9 (2013), 725–33.

Morus, Iwan Rhys, 'Placing Performance', *Isis*, 101 (2010), 775–8.

Nead, Lynda, *Victorian Babylon: People, Streets and Images in Nineteenth-Century London* (New Haven: Yale University Press, 2000).

O'Connor, Ralph, *The Earth on Show: Fossils and the Poetics of Popular Science, 1802–1856* (Chicago: University of Chicago Press, 2007).

Schechner, Richard, *Performance Studies: An Introduction* (New York: Routledge, 2002).

Striff, Erin, ed. *Performance Studies* (Basingstoke: Palgrave Macmillan, 2003).

Webb, David, 'For Inns a Hint, for Routes a Chart: The Nineteenth-century London Guidebook', *London* Journal, 6.2 (1980), 207–14.

Willis, Martin, *Vision, Science and Literature, 1870–1920: Ocular Horizons* (London: Pickering and Chatto, 2011).

Withers, Charles W.J and David N. Livingstone, 'Thinking Geographically about Nineteenth-Century Science', in *Geographies of Nineteenth-Century Science* eds Livingstone and Withers (Chicago: University of Chicago Press, 2011), pp. 1–19.

CHAPTER 4

Of Hats and Scientific Laughter

Tiffany Watt Smith

Abstract Tiffany Watt Smith brings to our attention a relatively little known Victorian dramatic performance: chapeaugraphy, or comic business with hats. Analysing this commonplace but now forgotten music hall performance enables Watt Smith to ask what relationships exist between dramas designed to elicit laughter and psychological studies of laughter, which also worked extensively with hats. Watt Smith concludes that psychological practices were indebted to late Victorian theatrical culture, a counterintuitive claim considering the common perception of science as continually hardening its experimental practices. Theatre, Watt Smith argues, remained a vital part of scientific culture because it offered a different, and more layered, mode of investigation.

'Ladies and Gentlemen, I propose to entertain you for a short time this evening by presenting a number of sketches under numerous shaped hats'.[1]

It's 'as hard as stand-up comedy'.[2] This is how Caspar Addyman, developmental psychologist at the Babylab, Birkbeck College, University of London, describes trying to understand what makes infants laugh. He, and the parents who volunteer their babies to participate in his experiments,

T.W. Smith
Queen Mary University of London, UK

© The Editor(s) (if applicable) and The Author(s) 2016   59
M. Willis (ed.), *Staging Science*,
DOI 10.1057/978-1-137-49994-3_4

grin, gurn, and blow raspberries when instructed by a nearby monitor. Their aim: to elicit giggles from tiny experimental subjects, their efforts recorded on video and analysed later by impartial, trained observers. Do 1-year olds find a surprise funnier than a joke they know is coming? What percentage of 6-month olds giggle when tickled? And why are hats so funny? Still at the beginning of his research, Addyman hopes to understand the links between laughter and learning, and why babies are so good at recruiting us into their games.

There is a playfulness in Addyman's tone when he compares himself to a stand-up comedian, but it is a joke worth paying attention to. At the Babylab emotions are not simply the objects of study. Addyman and the other researchers who help with his experiments, are also, to use the phrase coined by sociologist Arlie Russell Hochschild, doing 'emotional labour'.[3] Their laboratory roles explicitly require them to control their own affective presence and its effects. They demand that the scientists *perform*. Walk through the basement laboratories at the Babylab, and you will see walls painted grey to create a calm, unstimulating environment. The researchers are careful neither to worry nor excite their subjects, in the hope of generating a 'neutral' mood against which the subsequent effects of each experiment can be measured. In turn, once the games begin, Addyman and his colleagues must pay attention to their own performances: like stand-ups, they produce an atmosphere of fun on demand. With one eye on the spectacle they are making of themselves, the other on their audience, they experience the doubled consciousness which only comes when we know we are performing. No wonder then, that Addyman renders his experience in the language of the stage, his scientific practice more like being a stand-up comic than imitating the dispassionate gaze of an fMRI scanner or other measuring machine.

When in 1959 the novelist C.P. Snow spoke of 'two cultures'—the sciences in one camp, the arts and humanities in the other—he was lamenting a relatively recent development in intellectual life.[4] As the chapters in this volume by Morus and Willis as well as this present contribution show, only a few decades earlier the lines running between theatrical and scientific practices were more readily acknowledged. Does the complaint of 'two cultures' blind us to the ways in which these connections still linger, even if in less obvious ways? Theatres might seem to have little to offer the apparently cold process of producing hard scientific facts. From Plato's *Republic* onwards, they have been portrayed as spaces of distortion, where audiences and actors trade in misperceptions and become prone to emotional excess: theatricality, in this view, is the antithesis of rational knowledge. But even a glimpse at Addyman's project reminds us that some scientific practice has quite a lot in common with acting. As Addyman himself suggests, 'experiments are a lot like theatre, both are one-removed from the thing itself, a kind of imitation of reality in order to find something out'.[5] Addyman is not alone in noticing this similarity. In 1993, the philosopher of science Robert P. Crease traced the life of scientific experiments in theatrical language—from their tentative rehearsals to the tedious repetitiveness of a long run.[6] Thinking about theatre, then, might help us understand something about the nature of scientific enquiry—at the least, the embodied and emotional aspects of laboratory practice, and the kinds of performances that go into the creation of scientific facts.[7]

However, the analogy between theatre and scientific experiments is only part of the story. To look at Addyman's research from the perspective of performance and theatre studies reveals a number of shared techniques. Puppets, costumes, improvisation, and role-play are all part and parcel of life at the BabyLab. To invoke the theory of 'actor-networks' put forward by Bruno Latour, Andrew Pickering, and Karen Barad, these theatrical practices can be described as agents in scientific investigations, as being themselves 'actors' in the networks of objects, people, and processes which make it possible for scientific ideas to come into being.[8] Actor-network theorists typically speak of the *performativity* of laboratory technologies, invoking both the image of machines performing their function well (as in 'high performance'), and the idea, which J.L. Austin described, that words (or in actor-network theory, things and techniques) can perform, effecting some change in the world as we know and experience it.[9] I want to add to this talk of the performativity of laboratory techniques a further layer, and suggest that we might think about the theatricality of scientific practices too. Alongside laboratory codes and high-functioning equipment, theatre's guilty impostures and slippery pretences may also be construed as agents or 'actors', shaping our modern scientific understanding of the world.

This essay explores the relationship between scientific and theatrical practices in the late nineteenth and early twentieth centuries, a moment when the discipline of psychology is usually thought to have hardened into its modern form. It focuses on the links between the psychologist James Sully's studies of laughter, and the comic act Chapeaugraphy. Performed on variety stages at the *fin de siècle*, though little known today, Chapeaugraphy was a solo turn involving the manipulation of a piece of felt into variously shaped hats, under which the performer would strike comical poses. In these performances, two key motifs of this period's theatrical culture were brought together: first, a commitment to the comic potential of hats, and second, a fascination with transformation. This essay argues that these twin themes also animated Sully's scientific investigations into laughter, in which theatrical practices performed in various ways. Costumes, props, and clowning were part of the repertoire of techniques that Sully and his collaborators used to make laughter—and especially infant laughter—available for study: they were a way of gathering evidence, of *doing* things. But this essay also argues that the era's theatrical culture left its mark on how Sully and his scientific contemporaries *thought* about laughter too, informing the kinds of scenarios they took for granted as comic, and the examples with which they developed their scientific theories. As one might expect, echoes of these theatrical methods can be heard in the conclusions Sully drew. They reverberate in his vision of the laughing body as a site of perpetual transformation, and in turn, in his conviction that messiness is part of the way we perceive both humorous scenes and scientific facts.

An Essay on Laughter (1902)

Just over a 100 years ago, around the corner from the laboratories where Addyman makes babies laugh today, a group of philosophers and psychologists led by James Sully set up the UK's first fully equipped psychological laboratory at University College London. Opening its doors in 1898, the laboratory symbolized a new era in the study of mind. As the psychologist F.W.H. Myers put it, the new psychology would attempt 'to attack the great problems of our being not by metaphysical argument [...] but by a study, as detailed and exact as any other natural science'.[10] The anecdotal would be abandoned for the strictly observable, falsifiable, and repeatable experiments would take the place of ad-hoc tests, and machines would trump the fallible human gaze.

This modernizing rhetoric had a triumphant swagger, but in truth, studying the mind at the *fin de siècle* remained a complex and messy endeavour. As Roger Luckhurst, Jenny Bourne Taylor, and Martin Willis among others have shown, much of the new psychology's 'expert knowledge' had, in fact, 'uncertain provenance', and the approaches of the artist and amateur lingered alongside formal experimental regimens.[11] James Sully, as Grote Chair in Logic and Philosophy of Mind at University College London, played a key role in consolidating psychology into the professional discipline we know today. Yet, his own work also exemplifies the multiplicity of approaches which continued to define the field.[12] Sully had immersed himself in the philosophy of Alexander Bain and Herbert Spencer, and studied at the German psychophysical laboratories of Rudolf Hermann Lotze and Hermann von Helmholtz. He was an accomplished scientific writer, and known for his contributions to the era's lively and eclectic periodical culture, in particular his writing on music, literature, and drama for the *Fortnightly Review* and the *Cornhill Magazine*. It was perhaps a kind of wilful conviction that understanding the complexities of the human mind required more than a piece of laboratory technology could offer that impelled Sully, in 1902, to write a book-length study of laughter.

Laughing, giggling, guffawing, and sniggering were surprisingly important to Victorian men of science. The first decades of the nineteenth century had witnessed a transition in the language used to talk about our innermost feelings. The older theological concepts of the 'passions' and 'affectations of the soul' had begun to give way to a newly secular account of the 'emotions', in which the relationship between physiological reflexes and mental events was brought starkly into view.[13] Where in previous centuries, discussions of laughter had struck a moralistic tone—such as Hobbes's famous declaration that laughter comes from a feeling of 'sudden glory' when we compare our own triumphs to the failures of others—Victorian physiologists turned to the mechanical apparatus of the laughing body, and attempted to explain its spasms and contractions.[14] The key ideas underpinning this work can be traced back to the celebrated Victorian philosophers Herbert Spencer and Alexander Bain, both of whom argued that laughter was in essence a discharge of excessive 'nerve force' (their theory is today known as 'relief theory').[15] In *The Emotions and the Will* (1859), Bain argued that laughter erupts when we are released from the restraints of solemn or dignified behaviour, an unleashing of pent-up tension. Spencer disagreed, claiming that laughter was not caused by a release from constraint *per se*, but only when it involved what he called a 'descending incongruity'.[16] Imagine an audience at the theatre enrapt in a romantic drama, he wrote. The scene reaches its climax, the lovers are about to reunite. And then, quite unannounced, a young goat walks onto the stage, and begins to sniff at the actors. For Spencer, the audience's mirth at this theatrical misdemeanour was not simply a release from solemnity but the swift and total descent of the mind from a 'great' idea (the anticipation of lovers reunited) to a 'small' one (a lost goat). Spencer's explanation was enthusiastically discussed by

Charles Darwin in *The Expression of the Emotions in Man and Animals* (1872), who added to it his queries about the animal origins of sniggering, and reported variations in the outward demonstrations of amusement across the globe. Darwin also included reports of laughing lunatics from asylum superintendents, writing further that 'with idiots laughter is the most prevalent and frequent of all the emotional expressions'.[17] In this way, he gave scientific credibility to another impulse among late-Victorian authors, which regarded laughter as a peculiarly untrustworthy emotion, one common to criminals, the insane, and other 'degenerates'.[18] As the nineteenth century drew to a close, the riddle of laughter—how it helped, and when it imperilled—was firmly established as a key scientific question, tantalizing a number of intellectuals in Britain and the continent, including the novelist George Meredith, psychologists Théodule Armand Ribot, L. Dugas, and William McDougall, philosopher Henri Bergson, and psychoanalyst Sigmund Freud.[19] It is a testament to the importance of laughter at that time that in 1902, at the British Psychological Society's inaugural conference, Sully could give a paper entitled 'The Evolution of Laughter'.

Published that same year, Sully's book *An Essay on Laughter* has been characterized as primarily a synthesis resembling a modern 'literature review', its aim to 'organize and assess the relevant evidence' amassed by his contemporaries, without arguing for a 'particular approach'.[20] Yet, though it is true that much of his book is dedicated to existing debates, its guiding voice is relentlessly concerned with the idea that laughter itself has multiple causes. Laughter has a 'baffling spirit', wrote Sully.[21] It resists the many ingenious attempts made by nineteenth-century philosophers to 'twist [it] into a shape that will fit an adopted theory' (p. 8). There are laughs of triumph and of scorn; knowing smiles at clever word-play and sniggers at dirty jokes. Laughter shifts its form depending on culture and context, its expression dependent on gender and class and 'a thousand unknown influences of temperament and habits of life' (p. 20). It is this emphasis on what Sully called a 'plurality of causes', and his conviction that a variety of epistemic approaches were required to investigate them, which stands as his contribution to the period's scientific debate on laughter.

Unwilling to reduce laughter to a single unifying explanation as his predecessors had done, Sully set the latest physiological and evolutionary accounts alongside insights from novelists and philosophers and observations about giggling children by non-professionals. The effect is an occasionally disorientating patchwork of different methods and perspectives.

Lorraine Daston and Peter Galison have argued that in the late nineteenth century with the professionalization and separation of the scientific disciplines, mechanical objectivity—a gaze modelled on a dispassionate machine—was framed as a chief 'epistemic virtue'.[22] Sully's more eclectic approach hints at an alternative reading, one which suggests that the 'virtue' of mechanic objectivity, even at this key moment in the forging of a modern scientific practice, may have been more fraught than Daston and Galison allow. Sully was only too happy to admit the limitations of laboratory machines, writing that the physiology of laughter will never be fully understood until 'some ingenious experimenter can succeed in exciting the mirthful mood and at the same time cutting off the bodily reverberation [...]. It may be predicted with some confidence that this waiting will be a long one' (pp. 47–48). Yet even though his study incorporated expertise from beyond the laboratory, for Sully it remained a resolutely scientific project. Its starting point, he wrote, was the 'scientific pre-supposition' that laughter might have some underlying 'order and law' (p. 20)—even if that law might ultimately be the emotion's capacity for transformation. This search for underlying structures made his 'a serious enquiry into the subject' (p. 21). Sully's defence of his work as a 'serious' and 'scientific' enquiry reminds us that even for psychologists working in the early twentieth century, the ideal of a 'scientific' approach was both more capacious—and contested—than it might seem today.

As had his predecessors, Sully intuitively turned to the comic performances of the stage to communicate his ideas to his readers. From imagined scenes of funny-men performing for prehistoric audiences, to quotations from Shakespeare and Molière, theatre helps bring the writing in *An Essay on Laughter* to life. Yet, if we look closer, it is also possible to find theatrical performances animating Sully's work on laughter long before he set pen to paper. As it is for Addyman and his colleagues, role-playing and other kinds of pretence were important experimental resources for Sully, in particular in his observations of that evasive phenomenon, child laughter. The 'new child-study', wrote Sully, 'has not yet produced a methodological record of the changes which this interesting expression of feeling undergoes' (p. 200). In the absence of data and precise records, Sully drew on the diaries and letters which he had gathered for *Studies of Childhood* (1898).[23] The observations in this earlier book were based mostly on testimonies by parents and nursery workers who responded to a call Sully published in the journal *Mind* in 1893 requesting information about the infants in their care.[24] The reports Sully received made it abundantly clear that babies and children were not the only ones capable of 'roguish' pretences (p. 201). Entering into a spirit of scientific investigation, parents and nursery workers conducted experiments that required a skill in theatrical make-believe too. The American child-psychologist Millicent Shinn, for instance, described the first genuine laugh of her niece Ruth. It occurred on day 118: she 'was excited by the sight of the mother making faces'. Two years later, Ruth was led into a darkened room where her uncle was pretending to be a bear: she was convulsed with laughter and 'the exhilaration of unreal alarm' when he 'sprang out from his dark hiding-place growling fearfully' (p. 200). Such games, familiar to anyone who has spent time with young children, were for Sully elevated to the status of scientific experiments, in some cases deliberately staged to acquire evidence. Having heard from a Dr L. Hill that 'a child of four will laugh on being tickled much more vigorously than one of two' (p. 188), Sully attempted to test the theory. As had Darwin, Sully made the most of his own children as experimental subjects. He reports that Hill is correct: even 'the mere threat to apply [a tickle] suffices to evoke the reaction' in the 4-year-old child. Self-consciously enacting 'the mere threat' again and again, Sully transformed, for that moment, the nursery into a kind of scientific theatre in which an experimental observer was also a comic, one eye on his own performance, the other on its effect on his tiny audience.

Given the importance of theatrical performance to Sully's experiments on laughter, it is unsurprising to find that he came to view theatre as one of the chief mechanisms through which this emotional expression had evolved. For Sully, both physiological and cultural processes were responsible for shaping the laughter reflexes into their various present forms, and he thought the stage a particularly important influence on both. 'It is certain that the educative lead of the artist has been at work from a very early stage of human development', he wrote. Laughter—like all emotions—assumed its 'fullness and complexity' (p. 190) through repetition and habit, perfecting itself 'by practice' (p. 188). Convinced that comic performance was as old as humans themselves, Sully believed watching 'primitive funny men' (p. 189) to be a kind of bodily training, creating what Marcel Mauss would later term a 'habitus'[25]: 'the finer and more methodical exercise of men's gift of laughter by these skilled choragi must have been a potent factor in its development' (p. 192). Sully's account of the evolving complexity of laughter over time betrays the prejudices of his age. He believed that the 'simple' laughter of so-called primitive societies and children still lingered in the popular theatre and its more 'vulgar' aspects, writing of the 'boistrous fun of the spectacle of a good beating, for which the lower savages have a quick sense, and which is standing dish at the circus' (p. 348). At the same time, these developmental hierarchies are not entirely stable in Sully's writing. They are haunted by the unnerving prospect that the transformations which laughing bodies have already undergone are still in progress—and that even the most apparently involuntary reflexes can be reshaped under the influence of the untrustworthy charades of the stage.

The roles that theatrical techniques perform in Sully's *An Essay on Laughter* can be traced from experimental resource to evolutionary hypothesis. However, theatre's influence is felt in Sully's text in less obvious ways too. Hats might seem an unlikely starting point for an examination of the intersections of performance and science in early twentieth century psychology. Yet, they were a shared resource, playing important roles both on the comic variety stage and in scientific writing of this period.[26] In Sully's *Essay*, hats make repeated appearances as examples to think with: they mischievously run away with the wind, causing their owners to chase them down the street; they are accidentally sat on and crumpled; they are worn by the wrong people, or else block people's views at the opera with their absurd decorations. These hat-based comic scenarios, which Sully uses to develop his arguments, did not become funny on their own. The fact that they could be evoked as an example of the universally comic at all, is evidence of the deep interconnectedness of Edwardian scientific and theatrical practices. The reason hats were so helpful to Sully and his scientific contemporaries, I want to argue, is because they were already established tropes on the Victorian comic stage.

COMIC BUSINESS WITH *CHAPEAU*

The comic art of Chapeaugraphy is almost entirely forgotten today. The French clown Tabarin, who performed for crowds on the streets of eighteenth-century Paris is usually thought to have originated it. He carried a large floppy felt hat with the crown cut out, astonishing and entertaining passers-by by folding and twisting the felt into various shapes and striking humorous poses beneath them.[27] The technique was brought to England in the mid-nineteenth century by W.S. Woodin, a quick-change comic monologuist, whose 'rapid exchange of hats for caps, and caps for hats' was widely acclaimed.[28] However, it was the magician and entertainer Trewey, famous on the Continent for his Shadowography,

Ventriloquism, and Chapeaugraphy acts, who really brought the *chapeau* to the attention of British theatre-goers in the 1890s. Imitators quickly followed, portraying priests and Irishmen, nuns and Napoleons, all with a few swift twists of the felt. 'Capital exhibitions' of Chapeaugraphy are described in reviews of touring Christmas pantomimes; bills for variety shows called attention to 'marvellous' Chapeaugraphy interludes.[29] If not a headline act, then Chapeaugraphy was certainly a curiosity.[30] Catering to the bourgeois taste for amateur theatricals in late Victorian England, at least two instruction manuals were published, and a 'complete fit-out for performing Chapeaugraphy' at home went on sale at Hamley's Grand Magical Saloons in London—it included the *chapeau*, face chalks, the 'New Invisible Nose', and a moustache, all in a 'nice handy box, convenient for carrying about'.[31]

In his 1909 guide book, the magician and performer R.D. Chater, whose stage-name was Hercat, described how Chapeaugraphy should be performed. After introducing himself, the performer—wearing evening dress—should turn away from the audience and, with the aid of a mirror, quickly adjust the *chapeau* into the desired shape and add any additional props, such as a false nose, or a crucifix. Then he should spin back round for the 'reveal', holding the pose for a (surprisingly long) count of 12, before beginning the cycle again. Some caricatures required greater skill than others. 'The Cowboy' (see Fig. 4.1) was relatively straightforward: 'throw the ring over your head without any twist… but move it on one side, to represent a sombrero. Attach a moustache to your lip; tie a white handkerchief loosely round your neck, and carry a small coil of line in your hand to represent a lasso. Gaze over the heads of the audience as if you were scanning the horizon'. Others demanded a more sophisticated command of folding and twisting. Twenty or so 'portraits' would make up the turn: 'Beadle', 'Dustman' 'Coquette', 'Dunce', 'Magician', and 'Drunken Sailor' among them. Offensive national stereotypes were popular, particularly 'The Chinaman', 'The French Priest', and 'The Irishman' ('hold a short stick over your head. Assume an "Arrah, be jabbers!" expression'). The *dénouement* of this dizzying array of transformations was one final reveal:

> FINIS: 'Drop your cloak; thrust your head through the ring, allowing it to fall on your shoulders; bow to the audience and say, "Myself"'. (see Fig. 4.2)[32]

Some might find it hard to understand the appeal of Chapeaugraphy today. But enthusiastic Edwardian audiences described an 'amusing and

FIG. 8

FIG. 9

FIG. 10

Fig. 4.1 Hercat, *Chapeaugraphy, Shadowgraphy, and Paper-Folding* (London: Fleet Street, 1909). In the collection of the author

FIG. 29

FIG. 30

FIG. 31

Fig. 4.2   Hercat, *Chapeaugraphy, Shadowgraphy, and Paper-Folding* (London: Fleet Street, 1909). In the collection of the author

pleasing' diversion, and spoke of their astonishment at its 'marvellous' effects and 'bewildering' transformations.[33] Perhaps part of Chapeaugraphy's success can be explained by the fact that for Victorians and Edwardians, hats were already inscribed with layers of comic meaning. From Joe Gargery's painful attempt to place his hat on the mantelpiece in *Great Expectations* to Pooter's excruciatingly out-of-date straw boater in *The Diary of a Nobody*, hats were a favourite device in amusing scenes. They could differentiate a dustman from a drunkard, a gentleman from a sailor, quickly establishing the caricatures on which so much comic pleasure depends. There were also extravagant—and much-criticized and lampooned—millinery fashions, such as the vogue, in the 1880s and 1890s, for bedecking one's head with exotic plumage and even sometimes the taxidermied birds themselves.[34] Most of all, hats were part of a repertoire of formalized performances, which required a knowledge of when they should be doffed, removed, or replaced. Hats, then, were things with which mistakes could be made, and which could point up their owners' foibles or expose their pretensions.

Under the specific conditions of theatrical performance, however, hats did more than signify failures of etiquette. Perhaps more than any other object on stage, hats were also highly recognizable co-performers in comic routines. As Andrew Sofer has put it, objects on stage are not perceived by spectators as objects *per se*, but *as props*: they become a special sort of sign, achieving what he calls 'a set of semiotic quotation marks, so that a table becomes a "table"'.[35] The 'special sort of sign' that hats became in Victorian performance was freighted with a complex network of meanings—meanings that also included the sorts of things hats on stage were expected to *do*. Victorian and Edwardian audiences would have been extremely adept at recognizing this language and anticipating it too, since hat-based gags and routines were a very familiar part of their theatrical culture. Take, for instance, the diminutive music hall star Little Tich's most famous routine 'The Big Boot Dance' performed in the 1890s and 1900s.[36] The act turned on Tich's astonishingly dextrous attempts to keep his hat on his head (and therefore his clownish dignity intact), whilst wearing an oversized pair of boots which would sweep the hat into the air, or kick it away as he reached to pick it up. As J.B. Priestley recalled after Little Tich's death, during these routines the performer 'would suddenly take us behind the scenes with him, doing it with a single remark…He would drop a hat and be unable to pick it up because he kicked it out of reach every time and then mutter, half in despair, "Comic business with *chapeau*"'.[37] Tich's moment 'behind the scenes' gave the audience a *frisson* of complicity: it worked because the performer knew the audience would recognize that a predictable bit of comic business was underway. His audience knew to 'read' the hat not just as an object imbued with social status, but as the funny man to Tich's straight one, performing its part in a highly codified comic routine. Chapeaugraphy, then, might have been amusing—at least to some audience members— because it channelled a key principle of Edwardian popular performance practice: hats were capable of doing funny things.

If Chapeaugraphy was enjoyed for its hat-based gags, it also embodied another important preoccupation of Victorian and Edwardian theatrical culture—its fascination with magical mutations and quick changes. As Matthew Solomon has described, Chapeaugraphy grew out of a theatrical culture already 'characterized by transformations of various kinds'.[38] There were quick-change artistes, such as the monologuist W.S. Woodin alluded to earlier, who through rapid costume changes and impressive displays of voice, facial expressions, and posture conjured hundreds of different characters in a single show, earning him the sobriquet 'a talking Proteus'.[39] There were Shadowgraphers who, working behind a backlit screen, made rabbits morph into buildings and boats into trees. On a larger scale, in grand opera and theatre productions, money was poured into the kind of complex engineering tricks which could make Rhinemaidens float, or ghosts appear, buildings collapse or statues speak, all part of a stage culture which blurred the lines between scientific and theatrical spectacles.[40] Even whilst presenting itself as a nun's habit or cowboy's sombrero, the *chapeau* was haunted by the characters it had previously brought to life and those it was about to. The piece of felt was intriguing—exotic even— for its refusal to settle into one shape (perhaps this is one reason why its French name lingered even though, as Hercat pointed out, the phrase 'Chapeaugraphy', or 'writing in hats' was clumsy to say the least)[41]: the French *chapeau* conjured an air of mystery about the object, as if although it 'wrote' characters on stage, the felt itself was less easily legible. It was a 'thing' to evoke Bill Brown's discussion of this word in his essay 'Thing Theory', existing both in the material realm and 'beyond the grid of intelligibility'.[42] The clown Tabarin had described his *chapeau* as 'true raw material, indifferent to all forms [...] having almost no essential form except formlessness [...] for there is no moment, no instant, when it does not receive a new figure'.[43]

In the late Victorian and Edwardian theatre's fascination with ambiguity and mutability, it is possible to hear the echoes of that period's scientific interest in evolutionary theory.[44] Darwin had called the ideas of origins and originals into question, and proposed instead a vision of humans and animals in perpetual, if painstakingly slow, transformation. The world reimagined by evolutionary theory was one in which, as Darwin put it in the final sentence of *On the Origin of Species*, 'endless forms most beautiful and most wonderful have been, *and are being*, evolved' (my italics).[45] Yet, though the stage was surely influenced by a prevailing mood based in the natural sciences, theatrical transformations also found their way back into scientific writing, evidence of the dynamic exchange between the 'two cultures' in this period. As I have already described, stage practices were one of the 'actors' in Sully's experiments, bringing empirical facts about laughter into being. However, performing hats enjoyed even greater agency in his writing, as the examples through which Sully could sort through the implications of this evidence and shape his ideas. Sully approached laughter as an endlessly mutating reflex, coming to believe that it stemmed from an encounter with a grotesque or ill-fitting form—such as the *chapeau* between caricatures might appear to be. Additionally, he argued that perception itself was first an experience of a blurry whole, which only later resolved itself into its constituent parts. This understanding of a perceptual process characterized by mess and mutations served as a defence of Sully's own scientific approach to laughter which was, above all, guided by his commitment to multiplicity, and attraction to the in-between.

HAT BUSINESS AND THE SCIENCE OF LAUGHTER

The spectacle of a gentleman running after a top hat which had blown away in the wind. The case of 'a man sitting down on his own tall hat'.[46] The inherent absurdity of bulbous bowler.[47] From Spencer to Bergson, George Meredith to Freud, hats and their various misdemeanours were a recurring theme in scientific writing, and Sully's *An Essay on Laughter* is no exception. A 'flying hat pursued by its owner', the loss of one's hat, the sight of a man battering in an old hat, once-fashionable hats worn by unfashionable people, the 'glorious absurdity' of a 'hat and its bobbings' at a concert: each by turn, is proffered as an example of the universally comic.[48] Hats were so important in Sully's account of laughter that he even based a test for a sense of humour on one. 'The daily papers have not yet succeeded in inventing a satisfactory one, and the psychological laboratories have, wisely perhaps, avoided the problem', he wryly notes. But the following question, writes Sully, should suffice for predicting a person's capacity for fun: 'Supposing you made a call, and having placed your hat on a chair, inadvertently proceeded to sit on it; how would you feel?'.[49]

As Michael Billig has pointed out, the Victorian and Edwardian authors who wrote about laughter took for granted the idea that these hat-based comic moments were universally funny.[50] Common sense tells us that they were, almost certainly, not. Imagine being the maid who, forgetting to ask for the hat at the door, was responsible for its landing on the chair; or the nurse's fury at sticky fingerprints over a brand new silk hat. This blind-spot, however, does more than remind us of the ways scientific writing in this period reflected its authors' privileged positions. It also alerts us to the way the scenarios picked to illustrate the causes of laughter were also enmeshed in those deeply contrived comic routines so favoured by Victorian performers. Sully might not recognize it, but the hat-based scenarios which he proffers as examples of the inherently amusing had been moulded into their comic forms on the stage.

Although Victorian writers on laughter discussed a range of hat-based scenes, the sight of a child wearing a man's hat caused the most consternation. Child or 'midget' performers of the time are sometimes depicted wearing impressively large top hats, so it seems a fair assumption that this sight-gag was popular on the variety stage. However, it was also a familiar example discussed at length in scientific tracts on laughter, so often that Sully called it 'our old friend the child in the father's hat'.[51] In *An Essay on Laughter* eight pages are dedicated to trying to discern why this spectacle should provoke such hilarity.[52] The example becomes a resource to think about the processes of comic perception with, and one which ultimately allows Sully to defend his own practice of scientific study.

A common theory advanced at that time, was that seeing a child wearing a man's hat was funny because it involved a sudden and giddy descent from an important idea to an insignificant one. The idea of 'descending incongruity' which Spencer had suggested continued to play an important part in debates on the physiology of laughter. In 1898, the psychologist Theodore Lipps, made a related claim arguing that laugher is the result of a *Vorstellungsbewegung* or 'mental movement' between two ideas: first comes an idea of significance, next an idea of insignificance—or, as Sully translates it, first comes the 'belittling' idea and next the 'belittled' one.[53] Both Spencer and Lipps's arguments depend on the assumption that perception is a two-stroke motor: first we perceive the hat (the 'belittling' idea) and expect to see a man beneath it; second we see the child (the 'belittled' one). It is the effect of this sudden descent from important to unimportant which, Lipps supposed, would release the tension that the body expelled as laughter.

Sully used this same example to suggest a radically different account of comic perception. To do so, he re-staged the example of the child in the man's hat, imagining not a simple sight-gag based on incongruity, but a 'comic scene in the nursery' with a carefully orchestrated reveal: 'Let us suppose that a child in his nursery puts on his father's hat and stands on a chair, and that you enter the room and catch a glimpse of the hat first, say above a piece of furniture, and for a brief moment expect to see an adult beneath'.[54] Is the child in the man's hat funnier when he suddenly reveals himself? Or is it funnier to see the hat and the child all at once? For Sully, the element of surprise notwithstanding, both scenarios are equally comic. It is for this reason, he concludes, that laughter is not created by a sequential 'mental movement' from an important to unimportant idea, but by something altogether stranger.

'The newer psychology', writes Sully, 'teaches that in the first movement of perceiving an object we obtain not a distinct apprehension of parts, but a vague apprehension of a whole into which detail and definiteness only come later and gradually'. For Sully, to see the child wearing an adult's hat is not to see a hat and imagine a 'pre-existing pictorial type' and then compare this idea with the actuality. Instead, we laugh at the boy in the hat because we perceive 'a grotesque whole, *viz.* a hat on the wrong head'. The difference might seem subtle, but for Sully an important point was at stake. He believed that it was possible to recognize something as misplaced without having already brought to mind its correct position: 'when, after my servant has dusted my books and rearranged them on the shelves, I instantly recognize that they are wrongly placed, I may at that moment be quite unable to say what the right arrangement was'. Likewise, Sully believed it is possible to find an improperly worn hat funny without having already imagined it being properly worn. The child wearing a man's hat is funny not because we experience a deflating 'mental movement' from expectation to reality, but because we see something we instantly recognize as 'grotesque' and bizarre, even if we do not immediately know—or care—why. 'What tickles us', Sully concluded, 'is the uncustomary and topsy-turvy arrangement of *things*' (my italics).[55]

Through the pages of Sully's book, the laughing body emerges as itself an unstable form. Flux is an essential part of laughter's 'lightness and capriciousness of movement, [its] swift unpredictable coming and going', and its evolution in the play of historical as well as biological forces.[56] Additionally, for Sully, an experience of 'grotesque' or ill-fitting forms was the prompt for laughter, resulting not from a comparison with an ideal or correct arrangement, but because of an immediate sensation of unruliness. In many ways, the messiness which Sully saw as so integral to both laughter and the perception of the comic is also reflected in the form of his scientific investigations. In its relentless evocation of different kinds of evidence and juxtapositions of sometimes competing theories, *An Essay on Laughter* takes its readers on sometimes dizzying journeys through its pages. Like the boy in the man's hat, Sully's book might itself be thought of as 'an ensemble, which can only be described as a whole made up of ill-fitting parts'.[57] Its author searched for a quality of explanation quite different from the single, all-encompassing law which Bain or Spencer hoped to find. By drawing on varying techniques and approaches, Sully encountered the topic first as a blurry multiplicity, out of which laws and principles gradually came into focus. Sully's discussion of the boy-in-the-hat routine, then, allows him not only to argue for a particular understanding of mirth, but also to make a larger point about the ways we come to know ourselves and the world around us. Here, then, Sully's scientific observation collides with a moment of theatrical spectatorship, in which the world is encountered not through its articulated parts or ideal shapes, but as a 'grotesque' of constantly mutating forms, which come briefly into focus before disappearing again.

CONCLUSION

When historians look back at the roots of today's discipline of psychology, it is to the late nineteenth and early twentieth century that they turn. The sciences of mind in this period are usually associated with the hardening of a modern research practice, and above all with the emergence of a new 'epistemological virtue' which held mechanic objectivity—an approach modelled on the disembodied gaze of a laboratory machine—as its guiding principle and key metaphor. Influential though these new laboratory practices were, it is important not to assume that all psychological investigations of this period turned exclusively to them. This chapter has explored the various ways in which psychological practices and writing in the early twentieth century remained indebted to the period's wider theatrical culture. In Sully's scientific studies of laughter, comic acting was a key experimental resource, one which made laughter available for scientific observation. Additionally, comic performances—particularly those involving hats—offered a storehouse of incidents and examples, through which Sully was able to think about the nature of laughter, imagine the processes through which the response had evolved, and most of all, scrutinize the perceptual patterns at the heart of comic pleasure. Theatre then, for Sully, was a vital part of how a scientist could approach a complex problem like laughter. He understood that studying our giggles and guffaws required a messier, more layered mode of investigation than the strict parameters of the psychophysical laboratory could allow. Where the resulting book might seem to bulge with multiple and overlapping methods and theories, its effect is not only to portray laughter as a slippery and endlessly mutating object. It also reminds us that scientific looking at that time involved unexpected allegiances and complicities. Sully's effort to understand laughter, then, bursts its own seams, as much an excursion into the methods by which we come to know the world around us, as a discussion about a bodily reflex. The techniques and motifs of the stage played a key role in Sully's messier epistemological approach, their presence alerting us to theatre's influence in the embodied networks through which scientific knowledge was produced in the late nineteenth and early twentieth centuries. This period's psychologists were deeply indebted to theatre. And if as scholars, we look more closely at contemporary psychological practices, we might find that they still are.

Notes

1. Hercat, *Chapeaugraphy, Shadowgraphy and Paper-Folding* (London: Dean and Sons, 1909), p. 12.
2. Caspar Addyman and Isabel Addyman, 'The Science of Baby Laughter', *Comedy Studies*, 4.2 (2013), 143–153 (p. 144).
3. Arlie Russell Hochschild, *The Managed Heart: Commercialization of Human Feeling* (Berkeley: University of California Press, 1983).
4. C.P. Snow, *The Two Cultures and the Scientific Revolution. The Rede Lecture 1959* (Cambridge: Cambridge University Press, 1959).
5. Interview with Caspar Addyman at Birkbeck Babylab, conducted by the author, 6 May 2014.
6. Robert P. Crease, *The Play of Nature, Experimentation as Performance* (Indianapolis: Indiana University Press, 1994). For a discussion of Crease see: Tiffany Watt Smith, 'Cardboard, Conjuring and "A Very Curious Experiment"', *Interdisciplinary Science Reviews*, 38.4 (2013), 306–320 (pp. 312–313).

7. For further exploration of this idea, see Tiffany Watt Smith, *On Flinching: Theatricality and Scientific Looking from Darwin to Shell Shock* (Oxford: Oxford University Press, 2014).

8. See Bruno Latour, *Science in Action: How to Follow Scientists and Engineers Through Society* (Cambridge, Harvard University Press, 1987); Andrew Pickering, *The Mangle of Practice: Time, Agency and Science* (Chicago: University of Chicago Press, 1995); Karen Barad, *Meeting The Universe Halfway: Quantum Physics and the Entanglement of Matter and Meaning* (Durham: Duke University Press, 1997).

9. J.L. Austin, *How To Do Things With Words: The William James Lectures delivered at Harvard University in 1955,* eds J.O. Urmson and Marina Sbisa (Oxford: Clarendon Press, 1962).

10. W.H. Myers, 'Human Personality in the Light of Hypnotic Suggestion', *Proceedings for the Society of Psychical Research*, 4 (1886), 1–24, (p. 1).

11. Roger Luckhurst, *The Invention of Telepathy, 1870–1901* (Oxford: Oxford University Press, 2002), p. 21. See also Jenny Bourne Taylor, 'Psychology at the *Fin de Siècle*', in G. Marshall ed., *The Cambridge Companion to the Fin de Siècle* (Cambridge: Cambridge University Press, 2007); Martin Willis, *Vision, Science and Literature, 1870–1920: Ocular Horizons* (London: Pickering and Chatto, 2011).

12. See Bourne Taylor, 'Psychology at the *Fin de Siècle*', p. 24; Sally Shuttleworth, *The Mind of the Child: Child Development in Literature, Science and Medicine*, 1840–1900, (Oxford: Oxford University Press, 2010), p. 288.

13. See Thomas Dixon, '"Emotion": The History of a Keyword in Crisis', *Emotion Review*, 4.4 (2012), 338–344.

14. Thomas Hobbes, *Leviathan*, ed. Richard Tuck (Cambridge: Cambridge University Press, 1991), p. 43.

15. For a discussion of Victorian 'Relief Theory', see: Michael Billig, *Laughter and Ridicule: Towards a Social Critique of Humour* (London: Sage, 2005), pp. 86–110.

16. Herbert Spencer, 'The Physiology of Laughter' in Herbert Spencer, *Essays: Scientific, Political and Speculative* (Second Series), (New York: D. Appleton, 1864), p. 116.

17. Charles Darwin, *The Expression of the Emotions in Man and Animals* (London: Murray, 1872), p. 199.

18. For an extreme example of Victorian suspiciousness towards laughter, see: George Vasey, *The Philosophy of Laughter and Smiling* (London: J. Burns, 1875). For more on the pathologisation of laughter in this period see: Mackenzie A. Bartlett, *Laughing to Excess: Gothic Fiction and the Pathologisation of Laughter in Late Victorian Britain*, Unpublished PhD Thesis, (Birkbeck College, University of London, 2009).

19. T.A. Ribot, *The Psychology of the Emotions*, trans. H. Ellis (London: Scott, 1897); L. Dugas, *Psychology du Rire* (Paris: Félix Alcan, 1902); William

McDougall, 'A New Theory of Laughter', talk given to the *British Association* (1913); George Meredith, *An Essay on Comedy* (London: Archibald Constable, 1897); Henri Bergson, *Laughter: An Essay on the Meaning of the Comic*, trans. C. Brereton and F. Rothwell, (London: Macmillan, 1900, 1911); Sigmund Freud, *Jokes and their Relation to the Unconscious* trans. James Strachey (London: Hogarth, 1909, 1960).

20. Billig, *Laughter and Ridicule*, p. 105.
21. James Sully, *An Essay on Laughter* (London: Longmans, 1902), p. 4.
22. Lorraine Daston and Peter Galison, 'The Image of Objectivity', *Representations*, 40 (1992), 81–128 (p. 82).
23. James Sully, *Studies of Childhood* (London: Longmans, 1896).
24. See Shuttleworth, *The Mind of the Child*, p. 278.
25. Marcel Mauss, 'Techniques of the Body', trans. Ben Brewster, repr. in *The Body: A Reader*, ed. by M. Fraser and M. Greco, (London: Routledge, 1936, 2005), pp. 73–76 (p. 74).
26. I owe this latter observation to Billig, *Laughter and Ridicule*, pp. 107–108.
27. Hercat, *Chapeaugraphy*, p. 2.
28. Anon., 'Mr W.S. Woodin', *The Players, A Dramatic Musical and Literary Journal*, 3, 9 February 1861, p. 59.
29. Anon., *The Era*, 3202, 10 February 1900, p. 10.
30. The word 'chapeaugraphy' may not have been widely-known. When Hamley asked people if they knew of any other books written on the subject, 'they looked with astonishment at me and said that they had never even heard of it'. J.G. Hamley, *Chapeaugraphy or Twenty-Five Heads Under One Hat* (London: W. & F. Hamley, 1885), p. 2.
31. The two manuals are: Hamley, *Chapeaugraphy*; and Hercat, *Chapeaugraphy*. The 'fit out' is advertised in the back matter of Hamley, *Chapeaugraphy*.
32. Hercat, *Chapeaugraphy*, p. 16, p. 26, p. 28.
33. Hercat, *Chapeaugraphy*, p. 2.
34. Laurel T. Ulrich et al, *Tangible Things: Making History Through Objects*, (Oxford: Oxford University Press, 2015), p. 144.
35. Andrew Sofer, *The Stage Life of Props* (Ann Arbor, University of Michigan Press, 2003), p. 31.
36. The 'Big Boot Dance' was filmed in Paris in 1900, and released in 1903. Film: *Little Tich et ses 'Big Boots'*, dir. C. Gratioulet (Paris: Phono-Cinéma-Théâtre, 1903).
37. Quoted in Oliver Double, *Getting The Joke: The Inner Workings of Stand-Up Comedy* (London: Methuen, 2005), p. 177.
38. M. Solomon, '"Twenty-Five Heads under One Hat" in the 1890s', in *MetaMorphing: Visual Transformation and the Culture of Quick-Change*, ed. Vivian Sobchack (Minneapolis: University of Minnesota Press, 2000), pp. 3–20 (p. 4).

39. Anon., 'The Theatres &c.', *Illustrated London News*, 12 June 1859, p. 563.
40. For more on the role of scientific spectacles in theatrical performance see: Iwan Rhys Morus, 'Worlds of Wonder: Sensation and the Victorian Scientific Performance', *Isis*, 101 (2010), 806–816; Bernard Lightman, *Victorian Popularizers of Science* (Chicago: University of Chicago Press, 2007), pp. 167–218.
41. 'Chapeaugraphy is a hybrid word, a mixture of French and Greek. I am not aware who originated it, but it appears to me that a better word might have been devised': Hercat, *Chapeaugraphy*, p. 2.
42. Bill Brown, 'Thing Theory', *Critical Inquiry*, 28.1 (2001), 1–22 (p. 5).
43. Quoted in Solomon, 'Twenty-Five Heads under One Hat", p. 9.
44. For more on the influence of evolutionary theory on the nineteenth- and early-twentieth-century stage, see: Jane Goodall, *Performance and Evolution in the Age of Darwin: Out of the Natural Order* (London: Routledge, 2002); Kirsten Shepherd-Barr, *Theatre and Evolution from Ibsen to Beckett* (New York: Columbia University Press, 2015).
45. Charles Darwin, *On The Origin of Species* (London: Murray, 1859), p. 491.
46. William McDougall, *An Outline of Psychology* (London: Methuen, 1923), p. 167.
47. See: Fred M. Robinson, 'The History and Significance of the Bowler Hat: Chaplin, Laurel and Hardy, Beckett, Magritte and Kundera', *Tri-Quarterly*, 66 (1986), 173–200.
48. Sully, *Essay on Laughter*, pp. 97-8, 96, 214, 274, 327.
49. Sully, *Essay on Laughter*, p. 426.
50. Billig, *Laughter*, p. 108.
51. Sully, *Essay on Laughter*, p. 137.
52. Sully, *Essay on Laughter*, pp. 9-17.
53. Sully, *Essay on Laughter*, p. 9.
54. Sully, *Essay on Laughter*, p. 12.
55. Sully, *Essay on Laughter*, pp. 14, 15, 13.
56. Sully, *Essay on Laughter*, p. 21.
57. Sully, *Essay on Laughter*, p. 13.

BIBLIOGRAPHY

Addyman, Caspar and Isabel Addyman, 'The Science of Baby Laughter', *Comedy Studies*, 4.2 (2013), 143–153.

Anon., *The Era*, 10 February 1900, No. 3202, p. 10.

Anon., 'The Theatres &c.', *Illustrated London News*, 12 June 1859, p. 563.

Anon., 'Mr W.S. Woodin', *The Players, A Dramatic Musical and Literary Journal*, 9 February 1861, No. 3, p. 59.

Austin, J.L., *How To Do Things With Words: The William James Lectures delivered at Harvard University in 1955,* eds J.O. Urmson and Marina Sbisa (Oxford: Clarendon Press, 1962).

Barad, Karen, *Meeting The Universe Halfway: Quantum Physics and the Entanglement of Matter and Meaning* (Durham: Duke University Press, 1997).

Bartlett, Mackenzie A., *Laughing to Excess: Gothic Fiction and the Pathologisation of Laughter in Late Victorian Britain,* Unpublished PhD Thesis, (Birkbeck College, University of London, 2009).

Bergson, Henri, *Laughter: An Essay on the Meaning of the Comic,* trans. C. Brereton and F. Rothwell, (London: Macmillan, 1900; 1911).

Billig, Michael, *Laughter and Ridicule: Towards a Social Critique of Humour* (London: Sage, 2005).

Brown, Bill, 'Thing Theory', *Critical Inquiry,* 28.1 (2001), 1–22.

Crease, Robert P., *The Play of Nature, Experimentation as Performance* (Indianapolis: Indiana University Press, 1994).

Darwin, Charles, *On The Origin of Species* (London: John Murray, 1859).

Darwin, Charles, *The Expression of the Emotions in Man and Animals* (London: John Murray, 1872).

Daston, Lorraine and Peter Galison, 'The Image of Objectivity', *Representations,* 40 (1992), 81–128.

Dixon, Thomas, '"Emotion": The History of a Keyword in Crisis', *Emotion Review,* 4.4 (2012), 338–344.

Double, Oliver, *Getting The Joke: The Inner Workings of Stand-Up Comedy* (London: Methuen, 2005).

Dugas, Ludovic, *Psychology du Rire* (Paris: Félix Alcan, 1902).

Fraser, Mariam and Monica Greco, eds *The Body: A Reader* (London: Routledge, 2005).

Freud, Sigmund, *Jokes and their Relation to the Unconscious* trans. James Strachey (London: Hogarth, 1900; 1960).

Goodall, Jane, *Performance and Evolution in the Age of Darwin: Out of the Natural Order* (London: Routledge, 2002).

Gratioulet Clément-Maurice, *Little Tich et ses 'Big Boots'* (Paris: Phono-Cinéma-Théâtre, 1903).

Hamley, John G., *Chapeaugraphy or Twenty-Five Heads Under One Hat* (London: W. & F. Hamley, 1885).

Hercat, *Chapeaugraphy, Shadowgraphy and Paper-Folding* (London: Dean and Sons, 1909).

Hobbes, Thomas (1991), *Leviathan,* ed. Richard Tuck (Cambridge: Cambridge University Press, 1991).

Hochschild, Arlie Russell, *The Managed Heart: Commercialization of Human Feeling* (Berkeley: University of California Press, 1983).

Latour, Bruno, *Science in Action: How to Follow Scientists and Engineers Through Society* (Cambridge, Harvard University Press, 1987).

Lightman, Bernard, *Victorian Popularizers of Science: Designing Nature for New Audiences* (Chicago and London: University of Chicago Press, 2007).

Luckhurst, Roger, *The Invention of Telepathy, 1870–1901* (Oxford: Oxford University Press, 2012).

Marshall, Gail, ed *The Cambridge Companion to the Fin de Siècle* (Cambridge: Cambridge University Press, 2007).

McDougall, William, *An Outline of Psychology* (London: Methuen, 1923).

Meredith, George, *An Essay on Comedy* (London: Archibald Constable, 1897).

Morus, Iwan Rhys, 'Worlds of Wonder: Sensation and the Victorian Scientific Performance', *Isis*, 101 (2010), 806–816.

Myers, W.H., 'Human Personality in the Light of Hypnotic Suggestion', *Proceedings for the Society of Psychical Research*, 4 (1886), 1–24.

Pickering, Andrew, *The Mangle of Practice: Time, Agency and Science* (Chicago: University of Chicago Press, 1995).

Ribot, Théodule-Armand, *The Psychology of the Emotions*, trans. H. Ellis (London: Scott, 1897).

Robinson, Fred Miller, 'The History and Significance of the Bowler Hat: Chaplin, Laurel and Hardy, Beckett, Magritte and Kundera', *Tri-Quarterly*, 66 (1986), 173–200.

Shepherd-Barr, Kirsten, *Theatre and Evolution from Ibsen to Beckett* (New York: Columbia University Press, 2015).

Shuttleworth, Sally, *The Mind of the Child: Child Development in Literature, Science and Medicine, 1840–1900*, (Oxford: Oxford University Press, 2010).

Snow, C.P., *The Two Cultures and the Scientific Revolution. The Rede Lecture 1959* (Cambridge: Cambridge University Press, 1959).

Sobchack, Vivian, *MetaMorphing: Visual Transformation and the Culture of Quick-Change* (Minneapolis: University of Minnesota Press, 2000).

Sofer, Andrew, *The Stage Life of Props* (Ann Arbor: University of Michigan Press, 2003).

Spencer, Herbert, *Essays: Scientific, Political and Speculative* (Second Series), (New York: D. Appleton, 1864).

Sully, James, *Studies of Childhood* (London: Longmans, 1896).

Sully, James, *An Essay on Laughter* (London: Longmans, 1902).

Ulrich, Laurel Thatcher et al, *Tangible Things: Making History Through Objects* (Oxford: Oxford University Press, 2015).

Vasey, George, *The Philosophy of Laughter and Smiling* (London: J. Burns, 1875).

Watt Smith, Tiffany, 'Cardboard, Conjuring and "A Very Curious Experiment"', *Interdisciplinary Science Reviews*, 38.4 (2013), 306–20.

Watt Smith, Tiffany, *On Flinching: Theatricality and Scientific Looking from Darwin to Shell Shock* (Oxford: Oxford University Press, 2014).

Willis, Martin, *Vision, Science and Literature, 1870–1920: Ocular Horizons* (London: Pickering and Chatto, 2011).

CHAPTER 5

'You Can't Make a Film About Mice Just by Going Out into a Meadow and Looking at Mice': Staging as Knowledge Production in Natural History Film-making

Jean-Baptiste Gouyon

Abstract Jean-Baptiste Gouyon interrogates one of the most fascinating and controversial aspects of natural film-making: the "faking" of footage that audiences presume to be gathered in natural and untouched environments but which were in fact staged by the film-makers. Focusing attention on the period from the 1970s to the 2000s, Gouyon reveals that such practices were very often new modes of scientific investigation, and were recorded in "making-of" documentaries that detailed the work undertaken by production teams and naturalists. In particular the close relationships that film-makers often formed with the animals they were recording gave new insights into the relationships between the animal and the human, and between different forms of epistemology in natural history.

Speaking in October 2013 at a literature festival in Cheltenham to promote his autobiography, wildlife cameraman Doug Allan, famed for his

J.-B. Gouyon
University College London, UK

© The Editor(s) (if applicable) and The Author(s) 2016
M. Willis (ed.), *Staging Science*,
DOI 10.1057/978-1-137-49994-3_5

83

work on such series as *The Blue Planet* (BBC 2000), *Planet Earth* (BBC 2007), or *The Frozen Planet* (BBC 2011), ensured some publicity for himself with the incendiary statement that the BBC fakes wildlife shots all the time.[1] Allan was referring to the controversy that followed the broadcasting in 2011 of the series *The Frozen Planet* (BBC 2011). It had then emerged that images of the birth of polar bears had been taken not in the wild, as the commentary may have led viewers to believe, but in a specially constructed den in a Dutch zoo. On the day after the row had erupted, the BBC issued a weak, half-apologetic statement confessing further faked footage in the series: a caterpillar was filmed freezing in a box and snowflakes were formed 'in a controlled environment'. However, the BBC reassured audiences, the series 'met expected editorial standards', and was not intended to 'mislead viewers'.[2] Doug Allan pointed out that the BBC's defence had not been bold enough:

> You can't make a film about mice just by going out into a meadow and looking at mice. You need to introduce them to a safely built set in which they will be happy. There's a lot of skill in doing that.[3]

Allan concluded that the BBC should have been 'proud' of the way in which it gathered footage, and should have made it obvious, rather than 'hiding the explanation on its website'.[4]

The artificiality, or constructedness, defended by Allan, stands in apparent contrast to the 'unrealisable fantasy' of observational realism, usually held to be the staple of wildlife documentary.[5] However, if wildlife films are approached as performative films, which 'perform the action they name', wildlife film-making does not appear any longer to be in rupture with the factual film-making tradition in which it belongs.[6] Rather it can be seen as an extension of this tradition's objective—the representation of reality—doubled with a heightened awareness of the necessary construction this aim entails, and with the will to acknowledge it.[7] But wildlife documentaries are a particular kind of performative film. Not only do they involve performances of the factuality of nature, but also, and primarily, performances of science. For, as Bruno Latour demonstrated, agreement upon the factuality of nature is a consequence of the scientific endeavour.[8] Performing nature thus involves performing science. And as Doug Allan pointed out, natural history film-making demands the deployment of a huge amount of knowledge (of animal behaviour and life cycle, of ecosystems and biotopes, for example) and technical skill. The disclosure of their techniques, Allan argued, for instance in 'making-of documentaries' (known as MODs, and often shown at the conclusion of natural history films, these short factual segments depict the methods employed to create natural history programmes for film or television), should reinforce rather than undermine film-makers' cognitive trustworthiness.

Indeed, 'strategies of disclosure and concealment' are at the core of the relationship between audiences and scientific performers, as Iwan Morus has shown in his chapter here and also in his study of Victorian scientific showmen who specialised in producing illusions that beguiled onlookers in order to educate them.[9] In particular, revealing one's technical ability, one's 'hands-on know-how' usually provided good support for claims to cognitive authority.[10] Similarly, when wildlife film-makers display their ingenuity at re-constructing nature through MODs, they lay claims about the validity of film-making as a mode of relation to nature and animals, in order to produce reliable knowledge of the natural world.

The disclosure of technical skills (as opposed to their concealment) points to the notion of witnessing, established as the cornerstone of experimental science in the seventeenth century.[11] Robert Boyle, one of the main proponents of this new approach to making science, insisted that experiments should be conducted in public, or at least in social spaces that allowed witnesses to see for themselves how things were done. Boyle also devised 'a literary technology of virtual witnessing'.[12] A combination of visual representations and texts, this was intended to multiply witnesses to an experiment by incorporating those people at a distance, in both time and space. Images were pivotal to that purpose: '[b]y virtue of the density of *circumstantial detail* that could be conveyed [...] they [images] imitated reality and gave the viewer a vivid impression of the experimental scene'.[13] MODs are similarly intended to allow for virtual witnessing. They show viewers how things were done, demonstrating film-makers' 'property of skill', their capacity to control nature so as to generate valuable knowledge of it.[14]

Although this chapter is neither about mice nor about the specific case of the series *The Frozen Planet* (BBC 2011), it is a discussion of how and what it means for the documentary process to involve the staging of nature. The main object of study for this chapter is the 2001 feature film *Winged Migration* (Perrin 2001), directed by Jacques Perrin, and the associated MOD, created and directed by O Barbé in 2002.[15] I put this material in relation to other nature films, especially the 1972 *The Flight of the Snow Geese* (Bartlett and Bartlett 1972), directed by Des Bartlett and

Jen Bartlett, for which no MOD was shot. In both cases, trained birds were used in order to obtain close-up images of birds in flight, which were thereafter inserted into the wildlife films. I shall seek to explore how film-makers dealt, both on screen and in the support material for each film, with the apparent contradiction between wildlife and trained performers.

Comparing the two films brings to the fore an important contrast between narrative structures. In the 1972 feature, *The Flight of the Snow Geese*, both tame and wild birds appear on screen, but trained birds and wild ones are kept explicitly separate in two distinct storylines, conducted side by side within the same filmic space. In the case of *Winged Migration*, a single storyline runs in the feature film, and all the birds eventually appear on screen as wild ones. The explicit identification of the trained birds is kept for the MOD, whose narrative revolves around their rearing, training, and filming. Some 30 years separate the two films, and this formal contrast between the two suggests that in this interval the claimed status of the film-object, and the meanings attached to the film-making process in relation to the production of knowledge significantly evolved.

From approximately the late 1960s to the early 2000s, wildlife film-making had evolved through a process of increased 'purification' leading to the production of purified representations of the working of nature, devoid from any sign of human involvement and animal labour.[16] These are what scholars have called the 'Blue-Chip documentaries', wildlife films with a high production value and typically depicting a nature stripped of any trace of human presence.[17] These representations of the natural world have been criticised as encouraging a vision of nature as something to be visually enjoyed rather than physically engaged with.[18] Yet the develop-ment of these purified representations of nature has been paralleled with the practice of producing and releasing MODs alongside them. These MODs depict film-making as what Bruno Latour has called 'a work of translation', and films as 'hybrids of nature and culture' that create the conditions for the formation of networks of humans and non-humans.[19] Indeed, MODs specifically emphasise film-makers' relational engagement with nature and animals when producing their documentaries. In turn, the resulting blue-chip documentaries provide audiences with the pos-sibility to engage with the natural world in a seemingly unmediated way. In wildlife MODs wildlife film-making is fashioned as a form of meaning-ful exchange; as the shared labour of humans and animals that enables the former to obtain better knowledge and understanding of the latter. The disclosure of the material means involved in the production of the films is an invitation to audiences to share, as witnesses, in this labour of knowledge production.

To put it another way, parallel to the development of the blue-chip documentary genre is that of the MOD genre. This joint development signals a change in the ethics informing natural history film-making, from hands-off to hands-on. Wildlife film-makers from the early 1900s to the late 1960s (all of whom were naturalists turned cameramen) prided themselves on their ability to obtain images of animals without intervention (thus maintaining the postulate of a genuine separation between nature and culture). By contrast, from the late-1960s wildlife film-makers tended to promote film-making as an epistemologically valid means of intervening in nature, which they achieved via MODs. This genre can be seen as a means for film-makers to invite viewers to give their consent to this interventionist approach to producing wildlife films. Analytically, this evolution of the culture of wildlife film-making makes it necessary to consider the documentary and its making-of as a single object of study, in order to make sense of both.

In addition, two kinds of performance appear worth considering here, that of animals and that of film-makers.[20] In his work on performances, Richard Schechner has suggested that performers are at the same time 'not themselves' and 'not not themselves'.[21] In this interval between the character and the performer a commentary can be inserted.[22] Schechner's conceptualisation is useful in thinking of nature films as objects of knowledge, as outcomes of practices of making that embody cognitive claims about how the world is and how it works.

In nature films, animals, often tamed or at least habituated to humans, are staged as wild in order to represent what is known about their species. In these performances, they are at the same time not-themselves, for example they stand in for, or act as, their wild brethren, and not-not-themselves, since they remain individual animals, albeit made to perform in front of the camera to represent their species. Film-makers, when they appear in a MOD, tend to present themselves as knowledge producers (not-themselves), but also as film-makers presenting themselves as knowledge producers. In the interval which both types of performance creates, a commentary is inserted on the material process of film-making, as a necessary artifice if the film is to stand as object of knowledge at all. In turn, a claim is made about film-makers' status, whose professionalism is emphasised by contrast with naturalists' status as amateurs. This latter claim can be compared to similar ones laid by life scientists at the end of the nineteenth century, when the pursuit of knowledge in fields like comparative anatomy, zoology, or taxonomy became professional careers 'patrolled by disciplined experts'.[23] In what follows I shall reflect on these ideas through an examination of how the meaning of 'staging' has shifted from the source of 'fakery' to being the necessary condition of knowledge production in relation to wildlife film-making.

WINGED MIGRATION: STAGING AS PRODUCTION OF KNOWLEDGE

An account of a year in the life of migratory birds, *Winged Migration* is shot so as to create in spectators the impression that they are part of the flock. Nothing is left in the frame that would prevent the illusion of an intimate relationship with the animals from crystallising. The spoken commentary is reduced to a minimum; most of the soundtrack is constructed of bird sounds accompanied by vocal music. Birds in flight are shot close-up, placing spectators in a position from where they can share the way birds see the world whilst migrating. As Burt has argued, this 'shared glance [...] suggests that we are looking from within nature and not at nature'.[24] By installing spectators in a visually mediated physical intimacy with the animals, the film brings audiences closer to nature and fosters a strong emotional engagement with it. This imaginary physical proximity materialises the natural historical knowledge of the birds' odyssey, making it a real part of the viewers' lived experience.

In order to achieve this 'spectacular effect', a thousand birds of nearly 40 different species were turned into disciplined 'avian actors', using the technique of imprinting pioneered by Konrad Lorenz.[25] The birds were then brought on location by plane or truck, along known flyways for their species on every continent, and filmed against dramatic backgrounds to construct a visual geography of bird migration. This process involves, for example, first gathering the eggs of pelicans (obtained at the Djoudj bird sanctuary in Senegal). These were then carried in in-flight incubators to the film director's estate in Normandy (France) and incubated there in specially designed facilities until hatching. The chicks were then imprinted on a couple of human care-takers, who would train them to respond to their call and to get used to being filmed whilst flying. Once the birds had reached the appropriate development stage, they were packed in wooden crates and flown back to the Senegalese Djoudj bird sanctuary, to be filmed with their native environment in the background. They were then taken to Kenya, so that shots of pelicans flying above the Savanna with Mount Kilimanjaro as a backdrop could be obtained. Overall the aim was to create a cinematic representation of pelicans' migration across the African continent, as plotted by ornithologists. Throughout *Winged Migration*, accepted knowledge of the natural phenomenon of bird migration is thus reconstructed via a complex staging involving transportation, care, training, and management.

In the resulting feature, close-ups and intimate shots of these trained birds mix indiscriminately with larger shots of wild birds. These were taken in the field by cameramen sent around the world, who had been specifically instructed to use the same cameras and 35 mm film gauge as the crews filming the imprinted birds. Although this made it very cumbersome for field cameramen, more used to shooting in 16 mm, in this way their footage of wild birds could easily be edited with those of trained ones, thus dissolving the wild/tame dichotomy. Certainly, viewers would not be able to distinguish between the two kinds of footage were it not for the MOD released alongside the main feature. In that documentary the tone is celebratory, and making the film is presented both as human achievement and as adventure:

> In July 1998 on the island of Skrudur, a wildlife sanctuary south of Iceland, our shooting begins. No fixed shots. Our cameras must be free to move at will. This means hauling masses of heavy equipment by hand. To reach this barren rock 30 minutes from the mainland, we have to cross the open sea. Our adventure has begun.[26]

The MOD provides every detail about how things were done, explicitly juxtaposing behind-the-scene sequences with the corresponding scenes from the film. Fully disclosing that trained birds were used and how they had been trained, explaining which equipment was necessary to film them, from specially designed ultra-lights to radio-guided cameras through to hot-air balloons, the MOD is a demonstration of the film-makers' property of skill, of their ingenuity at controlling nature:

> It takes a lot of research to reconcile the constraints of flying with the needs of film-making. There is not established technique for this kind of shooting. We had to design and build everything from scratch.[27]

However, looking at this documentary more closely, a specific kind of staging can be recognised. First, science is staged so as to accommodate film-making as a legitimate means of knowledge production. Second, the making of the film is staged so that it appears scientific. In an effort to avoid the use of trained birds lessening the film's value as a source of knowledge about the natural phenomenon of bird migration, the MOD puts to work various narratives in order to define film-making as part of the scientific enterprise.

STAGING FILM-MAKING AS SCIENCE MAKING

To begin with, the whole project worked with and investigated the techniques of physiologist and naturalist Konrad Lorenz. In *Winged Migration* Lorentz's theory of imprinting was submitted to new tests, as species never before tried for this technique, it is claimed, were successfully used:

> In the 1930s, as part of his work on animal behaviour, Konrad Lorenz developed the concept of imprinting. This great Austrian naturalist and Nobel physiology prize winner made himself the foster father of dozens of baby geese. In applying this concept to make a movie we were heading into unknown territory. Konrad Lorenz's imprinting technique was not known to work with species other than geese. Whereas we want to fly not just with geese but also with ducks, swans, pelicans, storks and cranes.[28]

The making of *Winged Migration* is thus presented as consolidating and expanding existing knowledge, and as such it appears to participate in the making of science. The narrative at this point is reinforced by an almost scientistic aesthetic. Technological imagery pervades the representation of the moment when the chicks to be imprinted hatch, and of their subsequent rearing. They come to the world out of numbered eggs, lying not in nests but in aseptic metal boxes, in a spotless environment. Chicks are then shown being fed by anonymous hands gloved in latex and manipulating syringes, very much like laboratory animals. The message is clear: the birds appearing in the film may be trained ones but they have been obtained through technologically mediated, scientifically informed means. They are, it may be argued, rational and objective birds.

Just as with laboratory animals, viewers of the MOD are not encouraged to conceive of these birds as subjects. Each one of them is a specimen, standing for a flock, a group, and their species. Throughout the MOD, as indeed in the feature film, nothing enables viewers to identify either individual birds or birds as individuals. Notably, whilst the care-takers may have named the animals they looked after (as is suggested in the MOD), these names remain unknown to viewers. Such prevention of individualisation keeps at bay the idea that these birds may be pets, which would remove them from the realm of wildlife and lessen their value as objects of knowledge.[29] The birds, despite their imprinting and training, must remain worthy as entities that can answer questions about the natural phenomenon of bird migration.

The most efficient way of demonstrating that trained birds and wild birds remain equally valid as sources of knowledge of bird migration is to have scientists using them as objects of knowledge. In order further to assert film-making as a part of the scientific enterprise, the MOD displays scientists visibly enrolled in that project. Thus, the film-director Jacques Perrin is staged as the spearhead of 'one of the largest "private" ornitho-logical networks the world has ever known'.[30] We witness him hosting a meeting with scientists at his home and talking with them as equals. And once the shooting has begun, some of the scientists joined him on the set to advise:

> Jean Dorst, Guy Jarry, and Francis Roux, all from the Paris Museum of Natural History, signed on as our scientific advisers. With their help, we picked the best species to imprint, the best migration routes to follow, and the biggest gathering places of wild birds around the world. Their help throughout the film would prove invaluable. [31]

Some of these bird experts, whilst on location, were also able to produce knowledge for themselves. One of them, Henri Weimerskirch, from the Centre d'Etudes Biologiques de Chizé (CNRS), worked as an adviser with the film crew which took images of pelicans in Senegal. During their stay in Africa, Weimerskirch and his research team studied the energy expen-diture of pelicans flying in 'V' formation, which led to a publication in *Nature*.[32] Using one of the film-crew's ultra-lights they flew alongside the imprinted pelicans and measured their heart rate, thus demonstrating that flight formation enable birds to save energy.[33] In 2003 when *Winged Migration* was released in the USA, Weimerskirch was interviewed for a piece in the *New York Times*, as an adviser for the film. He was described as a biologist who 'had been working with birds for 20 years, studying the energetics of their flight', and was said to have 'of course [...] never flown with them' during all this time. He was then quoted: 'It was incredible to be with the animal itself [...]. There, you can see exactly how it works'.[34]

This latter story exemplifies how science is staged in relation to the making of a wildlife film, in order to fashion film-making as a participator in the production of knowledge. Scientists advising on the set are shown as having access to perspectives they would not have been able to reach otherwise. The film-set thus becomes a place where genuine knowledge is produced. This interpretation is corroborated by the funding application the film's producers submitted to the European Environmental Fund. Describing the project, and how scientists would be involved in it, it reads:

> [t]he specific outlook of these scientists and their knowledge of the bird [sic] different behaviours allow us to have a better cinematographic approach. On their side, this film is a unique opportunity to carry out comparative behavioural studies and to deepen their knowledge of birds.[35]

To have participating scientists appearing to produce original knowledge of the natural world with the film's imprinted birds further dissolves the distinction between wild and tame, perhaps even making it irrelevant.

The dual staging at work in the MOD—that of film-making as a legitimate way of producing reliable knowledge of the natural world, and that of science so that it can accommodate film-making as an appropriate form of participation in the scientific endeavour—is rendered necessary by another kind of staging going on in the feature film. Trained birds were used to represent aspects of the behaviour of wild birds in order to reconstruct the natural phenomenon of bird migration on screen. In the next section, I want to think through this kind of performance by considering the earlier film *The Flight of the Snow Geese*, which depicted a similar bird migration, and for which birds imprinted on humans were similarly trained in order to get close-ups of birds in flight. A notable difference from *Winged Migration* is that no MOD was released alongside *The Flight of the Snow Geese*. Rather, the film-makers chose to interweave the staging of trained birds within their representations of wildlife.

THE EXPERT NATURALISTS

Des Bartlett and Jen Bartlett's 1972 documentary followed flocks of snow geese as they travelled from the tundra around Hudson Bay in the Canadian Arctic to the Mississippi Delta. It also depicts the adventures of the naturalist film-makers. Written and produced by Colin Willock for Anglia TV as part of their successful series *Survival*, it was shown in Britain as a Christmas Special on ITV on 26 December 1972. The Bartletts spent the 4 months of the Arctic summer on the birds' breeding ground. Following scientists' advice they settled next to a colony in the vicinity of the McConnell River, sleeping in tents at a study base which the University of Western Ontario maintained there. During this time they collected a dozen orphaned goslings. In the words of Colin Willock:

> The idea was that Des [Bartlett] should collect a number of snow goose gos-
> lings on the tundra, hand-rear them and band them with coloured rings so
> that they would be recognizable in flight. He would then bring them down
> the flyway with the wild skeins in order to film them flying free.[36]

Having imprinted the goslings, the Bartletts then, as was the case during the pre-production phase of *Winged Migration*, habituated the birds to being filmed. 'Right from the beginning,' Des Bartlett explained in a letter to his producer, 'we used the tame goslings to cut in with the wild action'.[37] Later, these tame birds enabled them to shoot slow-motion footage of geese in flight: 'Shot against a clear sky background, it gave the impression that you were actually flying wingtip to wingtip with the geese as part of their skein'.[38] Or, as the Bartletts commented in the film's accompanying book, which they published in 1975, these close-ups 'make the viewer feel that he [sic] is seeing the geese through the eyes of one of the birds actually in formation'.[39] Just like in *Winged Migration*, tame birds were thus used to obtain footage which would have been impossible to get otherwise. Similarly, they contributed to creating in viewers an impression of intimate physical proximity with the birds, the impression of being part of nature rather than simply spectators of it.

An important difference between the two films, however, is that two distinct storylines were constructed within *The Flight of the Snow Geese*, which kept separate each kind of birds. The first storyline was that of the flocks of snow geese migrating from the North to the South of the north-American continent which provided the main focus of the film. The second storyline was the story of the rescue, imprinting, training, and filming of orphaned goslings. As Colin Willock's account of the production of the film shows, this secondary storyline was built in within the film as the shooting was progressing and the rushes started to arrive in London. Willock and the Bartletts agreed that the story of the film-makers rescuing orphaned birds, looking after them, and filming them, 'the new thought which we both share enthusiastically: the question of the tame family of geese', would make for a compelling movie.[40] Within the 1 hour film, this interlaced narrative is one of the elements that invite audiences to trust the film-makers as reliable and trustworthy sources of knowledge, as it demonstrates their intimacy with the natural world.[41]

The Bartletts freely admitted in the companion book to having named the imprinted birds, which they called collectively 'the Creeps'. They also emphasised their ability to individualise the birds, which set them apart from 'strangers': 'to a stranger each of the Creeps looked the same, but we had no trouble telling them apart'.[42] In the film, the imprinted birds appear as pets that nonetheless provide the film-makers and audiences with insights about the natural world they could not obtain otherwise. As the Bartletts concluded, 'the Creeps taught us a great deal about the ways of snow geese'.[43] Through the 'superb-slow-motion footage', 'these miraculously beautiful shots' filmed whilst the imprinted birds were flying alongside the Bartletts' station wagon, viewers were offered knowledge of, for example, the anatomical aspects of bird flight.[44] These shots were edited to provide the film's opening sequence. Nevertheless, a later sequence revealed how they were obtained, thereby making sure that the images would not be mistaken for those of wild birds.

Maintaining the wild/tame boundary was essential for the film's cognitive legitimisation. The film-makers' demonstrated ability to travel safely across that boundary without misrepresenting nature stood as further evidence of their reliability as knowledge producers. The trust obtained by the film-makers in turn brings cognitive legitimacy to the film. In *Winged Migration*, by contrast, trust is solicited less for the film-maker than for film-making, which is actively presented as a mode of knowledge production in which the tame/wild distinction is irrelevant.

CLAIMING ARTIFICIALITY BACK

The comparison of these two films, produced three decades apart, suggests that in the intervening period a move took place within the culture of natural history film-making, away from the regime of cognitive legitimisation originating in the cultural repertoire of amateur natural history, and towards one pertaining to professional film-making. In 1971, in the introductory chapter of a book titled *Making Wildlife Movies*, Christopher Parsons, then a senior producer at the BBC Natural History Unit, wrote: 'the film-maker's only obligation to his audience is to ensure that his film is true to life, *within the accepted conventions of film-making*'.[45] As Parsons explained, his phrase in italics specifically refers to the use of captive or tame animals in place of wild ones in order to get shots that would be very difficult to obtain in the wild, 'yet would make a point of some value to the film'.[46] However, as Parsons also notes, 'Purists will throw up their hands in horror at such an idea'.[47] The 'purists' here are naturalists turned film-makers, who dominated natural history film-making from the early 1910s to the late 1960s before it became a profession.[48] To them, the essence of film-making laid in pitting their senses against the animals'.[49] Not only did they value the skills and virtues of field observation, 'field craft' and natural history's 'aesthetic of close detachment', they also appropriated those of big game hunters, displacing their authority as experts of the natural world.[50]

Cherry Kearton (1871–1940), a naturalist active before the First World War and in the interwar period, noted that naturalist cameramen would demonstrate their natural historical knowledge with footage that showed animals:

> not aware of anything near them, their behaviour being as unsuspicious as that of English cattle at a pond. It is this naturalness in a picture that is the test of the photographer's success. If a film shows animals alert, watch-ful, and suspicious, it is sure proof that the photographer was not properly hidden.[51]

These natural history film-makers were unarmed hunters, whose camera had replaced the gun. Each close-up of an animal was the result of 'a battle of wits'.[52] Their expertise rested on their capacity to endure the patience and discomfort involved in getting as close as possible to a wild animal without its knowledge. The phrase 'unarmed hunters' was used in 1963 as a title for a half-hour documentary depicting the film-making work conducted at the BBC Natural History Unit (NHU) in Bristol. The film, produced and directed by Christopher Parsons, affixes the label 'unarmed hunter' to more people than just the naturalist cameramen working in the field: including producers, film editors, sound recordists, and cameramen whose stock in trade was to film animals such as fish, amphibians, insects or small mammals in a controlled environment. *Unarmed Hunters* (BBC 1963) indicates that by the mid-1960s, the BBC NHU was starting to initiate a shift regarding what were to be accepted as appropriate practices in natural history film-making. Notably, the commitment to observational realism, an almost undisputed dogma until the 1970s amongst wildlife film-makers, made way in favour of a claimed artificiality. This shift was accompanied by the appearance of the MOD genre, which it is tempting to see as an off-shoot of the direct cinema movement; wildlife film-makers self-reflexively applying to themselves the claims of observational realism they were at the same time about to abandon in their own films. After *Unarmed Hunters*, Mick Rhodes produced a *Horizon* episode unambiguously titled 'The Making of a Natural History Film' (BBC 1972). As in 1963, it placed the emphasis on the technical work of wildlife film-making, notably disclosing how close-up shots of a wood was playing its eggs, or of tadpoles, could be obtained by filming these animals with macro cinematography equipment under controlled conditions.

These MODs all present film-making as a collective work, in contradistinction with the image of the solitary and heroic naturalist cameraman who prevailed in what could be called the pre-making-of era. They also insist that some forms of staging do not invalidate the films scientifically. Instead, they turn the film studio, where reconstructed bits of nature are used to shoot in a controlled environment, into a place akin to a laboratory where natural phenomena can be studied away from the vagaries of the field. Finally, these documentaries participate in the fashioning of professional wildlife film-makers as participants in the sciences, as they show them interacting with scientific practitioners on the set.

CONCLUSION: FACT AND FICTION

From *Winged Migration*'s MOD, the birds emerge also as actors. Their carefully staged performances made possible the production of genuine knowledge about the natural phenomenon of bird migration. The film-maker emerges as a demiurge, an active and subjective creator of meaning about the natural world. Not is the film-maker anymore soliciting trust as a 'silent watcher', as a passive and objective recorder of reality, but instead, through the expert deployment of film-making skills, is revealing aspects of the natural world that would have remained invisible within the boundaries of observational realism.[53] This suggests a shift in the values and meaning associated with staging and reconstruction. Instead of being seen as potential sources of fakery, they are claimed as a practical means towards a better understanding of the natural world. As much as it dissolves the wild/tame distinction, *Winged Migration* exemplifies a dissolving of the boundary between facts and fiction. The film medium loses its assumed transparency. Both the production and the reception stage gain in thickness as they become key sites of knowledge production. At the junction between these two sites stands the illusionary power of film.

As a reply to the charge of fakery brought against *The Frozen Planet*, David Attenborough (who narrated the British version of the series) offered the following defence:

> The question is, during the middle of this scene when you are trying to paint what it is like in the middle of winter at the pole, to say 'Oh, by the way, this was filmed in a zoo'. It ruins the atmosphere, and destroys the pleasure of the viewers and destroys the atmosphere you are trying to create [...]. Come on, we were making movies.[54]

Here Attenborough suggests that natural history film-making can use fiction as a means of producing facts. He argues in favour of the need to preserve the illusionary power of film, using it as an argument to defend wildlife film-makers' right to construction. With this conflation of the two supposedly separate realms of facts and fiction emerges the notion that imagination is an essential resource for wildlife film-makers to draw on if their films are to work as objects of knowledge.

If we consider the history of the genre of the natural history film, the comparison between *The Flight of the Snow Geese* and *Winged Migration* suggests that during the late 1960s and early 1970s, natural history film-making moved away from the realm of amateur natural history and became a professional pursuit patrolled by technical experts. This evolution entailed a change in the way natural history film-makers would support their claims to knowledge. In the early decades of the twentieth century, they echoed Victorian amateur naturalists' and big game hunters' discourse, foregrounding such themes as patience, self-discipline, self-restraint, bodily suffering, communion with nature, and the ability to outwit animals. By contrast, late twentieth-century wildlife film-makers would rather highlight their mastery of the film-making apparatus, and all the tricks they used to recreate nature on screen. The 1970s is a period of transition between these two regimes, with self-styled professional natural history film-makers like the Bartletts demonstrating their technical expertise whilst remaining true to the traditional values and beliefs of amateur natural history. The appearance and development of the MOD seems to be a good indicator of this transition. Indeed, it arose from the necessity to remove from the films everything that could destroy the atmosphere film-makers were trying to create and thereby reduce viewers' pleasure. At the same time the MOD enables wildlife film-makers to claim back the artifice and define their practice as a meaningful way of intervening in nature, and thus of knowing it.

From a broader perspective, the reading offered in this chapter contributes to a wider understanding of scientific performance, or of the relations between science and performance. Disclosing the practical means involved in performances of science which, ultimately are demonstrations of performers' 'property of skill', MODs stand as evidence of film-makers' capacity to control nature so as to generate valuable knowledge of it.[55] But, this chapter shows, in addition to defining film-makers' personal identity as trustworthy spokespersons for nature, MODs are statements which characterise the material practice of film-making as relevant to the production of knowledge. This is, in significant ways, similar to the instrumental skills required by scientific demonstrators of the Victorian period, as Iwan Morus has shown in the opening chapter to this volume. In these documentaries science and film-making are both staged so that the latter can appear as a material practice that valuably contributes to science. At the juncture of this dual staging stands the intrinsic artifice of film-making. As a joint staging of science and film-making, the MOD institutes artifice as necessary to obtain knowledge of nature. Considering how masks in Noh drama, too small to completely dissimulate performers' visage, create an interval between actors and characters, Richard Schechner meditates that the essence of the performance, as a means of understanding, resides in this interval between the performer and the performed.[56] Likewise, nature on screen is at the same time not-nature and not-not-nature. And from the interval between the two, generated by the artifice of film-making, new understandings, new knowledge of nature can originate.

NOTES

1. Rosie Taylor, 'BBC "fakes wildlife shots all the time": Veteran cameraman claims species 'smaller than rabbits' are filmed on custom-built sets', *Daily Mail*, 9 October 2013.
2. Daily Mirror, 'Frozen Planet scandal: Sir David Attenborough defends fake polar bear footage', *Daily Mirror*, 13 December 2011.
3. Taylor, 'BBC "fakes wildlife shots all the time"'.
4. Taylor, 'BBC "fakes wildlife shots all the time"'.
5. Stella Bruzzi, *New Documentary* second edition (London: Routledge, 2006) p. 217.

6. Bruzzi, *New Documentary*, p. 187.
7. Bruzzi, *New Documentary*.
8. Bruno Latour, *We have never been modern* (Cambridge; Harvard University Press, 1993).
9. Iwan Rhys Morus, 'Seeing and Believing Science', *Isis*, 97 (2006), 101–110 (p. 105).
10. Iwan Rhys Morus, 'Worlds of Wonder. Sensation and the Victorian Scientific Performance', *Isis*, 101(2010), 806–816 (p. 807).
11. Steven Shapin and Simon Schaffer, *Leviathan and the Air-Pump* (Princeton; Princeton University Press, 1985).
12. Shapin and Schaffer, *Leviathan*, p. 60.
13. Shapin and Schaffer, *Leviathan*, p. 62—emphasis in the original.
14. Iwan Rhys Morus, 'Manufacturing Nature: Science, Technology and Victorian Consumer Culture', *The British Journal for the History of Science*, 29 (1996). 403–434 (p. 416). See also Timothy Boon, *Films of Fact* (London: Wallflower Press, 2008), especially pp. 26–27, for a related discussion of early science cinema as a witnessing of science.
15. O Barbé, *The Making-of Winged Migration* (Paris: Galatée Films, 2002).
16. Latour, *Never been modern*, p. 11.
17. See for instance Derek Bousé, *Wildlife Films* (Philadelphia: University of Pennsylvania Press, 2000); Simon Cottle, 'Producing Nature(s): On the Changing Production Ecology of Natural History TV', *Media, Culture and Society*, 26 (2004), 81–101.
18. Gregg Mitman, *Reel Nature* (Cambridge, MS: Harvard University Press, 1999), p. 206.
19. Latour, *Never been modern*, pp. 10–11.
20. To an extent, nature films appear as performative films, conceptualised by Stella Bruzzi as films that perform what they name, see Bruzzi, *New Documentary*.
21. Richard Schechner, *Between Theater and Anthropology* (Philadelphia: University of Pennsylvania Press, 1985), p. 6.
22. Schechner, *Between*, p. 9.
23. James A. Secord, 'The crisis of nature', in *Cultures of natural history* eds Nick Jardine, James Secord, and Emma C. Spary (Cambridge: Cambridge University Press, 1996), pp. 447–459 (p. 449).
24. Jonathan Burt, *Animals in films* (London: Reaktion Books, 2002), p. 47.
25. The notion of 'spectacular effect' is developed in Fred Nadis, *Wonder shows: performing science, magic, and religion in America* (New Brunswick: Rutgers University Press, 2005). The phrase 'avian actors' comes from Jacques Perrin and Jean-François Mongibeaux, *Winged Migration* (San Francisco: Chronicle Books; Paris: Editions du Seuil, 2003). Imprinting is a technique formalised by Austrian ethologist Konrad Lorenz and consist-

ing in habituating a bird to a single human being from birth so that the bird will direct its instinctive behaviour towards the human as if it were its parent. For more on Lorentz see Richard Burkhardt, *Patterns of Behaviour* (Chicago: University of Chicago Press, 2005).

26. Barbé, *The Making-of Winged Migration*.
27. Barbé, *The Making-of Winged Migration*.
28. Barbé, *The Making-of Winged Migration*.
29. Lynda Birke, 'On Keeping a Respectful Distance', in *Reinventing biology: Respect for life and the creation of knowledge* eds. L.I. Birke and R. Hubbard (Bloomington and Indianapolis: Indiana University Press, 1995), pp. 75–88.
30. Perrin and Mongibeaux, *Winged Migration*, p. 230.
31. Barbé, *The Making-of Winged Migration*.
32. Henri Weimerskirch & al., 'Energy saving in flight formation', *Nature*, 413 (2001), 697–698. It is notable that the postal address of three co-authors of this study is that of Jacques Perrin's production company, Galatee film, thus squarely installing the film project within the boundaries of science.
33. The conclusion of this study is mentioned in the 2009 BBC series *Life*, in the episode 'Birds', during the sequence showing pelicans flying in formation. The producer of the episode, Patrick Morris, confirmed that Weimerskirch's paper was actually used as a source for the commentary (Patrick Morris, personal communication to the author). It is a nice case of knowledge obtained during the shooting of a natural history film being then communicated through another nature film.
34. James Gorman, 'Inviting Humans to Sprout Wings and Soar', *The New York Times*, 15 April 2003.
35. Anonymous, 'The Winged Migration' (Brussels: European commission, 1999).
36. Colin Willock, *The World of Survival* (London: André Deutsch, 1978), p. 127.
37. Willock, *The World*, p. 131.
38. Willock, *The World*, p. 147.
39. Des Bartlett and Jen Bartlett, *The Flight of the Snow Geese* (Toronto and London: Collins and Harvill Press, 1975), p. 114.
40. Willock, *The World*, p. 135.
41. Jean-Baptiste Gouyon, 'From Kearton to Attenborough. Fashioning the telenaturalist's identity', *History of Science*, 49 (2011), 25–60.
42. Bartlett and Bartlett, *The Flight*, p. 116.
43. Bartlett and Bartlett, *The Flight*, p. 175.
44. Willock, *The World*, p. 133, and p. 146.

45. Christopher Parsons, *Making Wildlife Movies, an introduction* (Newton Abbot: David & Charles Ltd., 1971), p. 14—emphasis in the original.
46. Parsons, *Making*, p. 15.
47. Parsons, *Making*, p. 15.
48. Jean-Baptiste Gouyon, 'The BBC Natural History Unit: Instituting natural history film-making in Britain', *History of Science*, 49 (2011), 425–451.
49. Christopher Parsons, 'The Silent Watcher', in *Look: a selection from the BBC-TV natural history series* ed. J. Boswall (London: British Broadcasting Corporation, 1969), pp. 13–19 (p. 18).
50. The 'aesthetic of close detachment' is from Gail Davies, 'Narrating the Natural History Unit: institutional orderings and spatial strategies', *Geoforum*, 31 (2000), 539–551. For the genealogy of natural history film-making in big game hunting see Gouyon, 'From Kearton'.
51. Cherry Kearton, 'Big Game hunting. Sport with the camera', in 'Film Number', *The Times*, supplement to Issue No: 45155, 19th March 1929, p. x, quoted in Gouyon, 'From Kearton', 38.
52. Parsons, 'The Silent Watcher', p. 18.
53. Parsons, 'The Silent Watcher'.
54. Daily Mirror, 'Frozen Planet scandal: Sir David Attenborough defends fake polar bear footage', *Daily Mirror*, 13 December 2011.
55. Morus, 'Manufacturing Nature'. See also Morus, 'Worlds of Wonder'.
56. Schechner, *Between*, pp. 6–9.

BIBLIOGRAPHY

Anonymous, 'The Winged Migration' (Brussels: European commission, 1999), accessed via <http://www.ec.europa.eu/environment/funding/projects/1999/3a.htm>.
Barbé, Oli, *The Making-of Winged Migration* (Paris: Galatée Films, 2002).
Bartlett, Des and Bartlett, Jen, *The Flight of the Snow Geese* (Norwich: Anglia Television, 1972).
Bartlett, Des and Bartlett, Jen, *The Flight of the Snow Geese* (Toronto and London: Collins and Harvill Press, 1975).
BBC, *The Making of a Natural History Film* (London: BBC, 1972).
BBC, *The Blue Planet* (London: BBC, 2000).
BBC, *Planet Earth* (London: BBC, 2007).
BBC, *The Frozen Planet* (London: BBC, 2011).
Birke, Lynda, 'On Keeping a Respectful Distance', in *Reinventing biology: Respect for life and the creation of knowledge*, eds L.I. Birke and R. Hubbard (Bloomington and Indianapolis: Indiana University Press, 1995), pp. 75–88.
Boon, Timothy, *Films of Fact* (London: Wallflower Press, 2010).

Bousé, Derek, *Wildlife Films* (Philadelphia: University of Pennsylvania Press, 2000).

Bruzzi, Stella, *New Documentary*, second edition, (London: Routledge, 2006).

Burkhardt, Richard W., *Patterns of Behavior: Konrad Lorenz, Niko Tinbergen, and the Founding of Ethology* (Chicago: The University of Chicago Press, 2005).

Burt, Jonathan, *Animals in films* (London: Reaktion Books, 2002).

Cottle, Simon, 'Producing Nature(s): On the Changing Production Ecology of Natural History TV', *Media, Culture and Society*, 26 (2004), 81–101.

Daily Mirror, 'Frozen Planet scandal: Sir David Attenborough defends fake polar bear footage', *Daily Mirror* 13 December 2011, accessed via <http://www.mirror.co.uk/tv/tv-news/frozen-planet-scandal-sir-david-96593>.

Davies, Gail, 'Narrating the Natural History Unit: institutional orderings and spatial strategies', *Geoforum*, 31 (2000), 539–551.

Gorman, James, 'Inviting Humans to Sprout Wings and Soar', *The New York Times*, 15 April 2003, accessed via <http://www.nytimes.com/2003/04/15/science/inviting-humans-to-sprout-wings-and-soar.html>.

Gouyon, Jean-Baptiste, 'From Kearton to Attenborough. Fashioning the telenaturalist's identity', *History of Science*, 49 (2011), 25–60.

Gouyon, Jean-Baptiste, 'The BBC Natural History Unit: Instituting natural history film-making in Britain', *History of Science*, 49 (2011), 425–451.

Kearton, Cherry, 'Big Game hunting. Sport with the camera', in 'Film Number', *The Times*, supplement to Issue No: 45155, 19th March 1929, p. x.

Latour, Bruno, *We have never been modern* (Cambridge: Harvard University Press, 1993).

Mitman, Gregg, *Reel Nature* (Cambridge: Harvard University Press, 1999).

Morus, Iwan R., 'Manufacturing Nature: Science, Technology and Victorian Consumer Culture', *The British Journal for the History of Science*, 29 (1996), 403–434.

Morus, Iwan R., 'Seeing and Believing Science', *Isis*, 97 (2006), 101–110.

Morus, Iwan R., 'Worlds of Wonder. Sensation and the Victorian Scientific Performance', *Isis*, 101 (2010), 806–816.

Nadis, Fred, *Wonder shows: performing science, magic, and religion in America* (New Brunswick: Rutgers University Press, 2005).

Parsons, Christopher, *Unarmed Hunters*, (Bristol: BBC-Bristol, 1963).

Parsons, Christopher, 'The Silent Watcher', in *Look: a selection from the BBC-TV natural history series*, ed. J. Boswall (London: British Broadcasting Corporation, 1969), pp. 13–19.

Parsons, Christopher, *Making Wildlife Movies, an introduction* (Newton Abbot: David & Charles Ltd, 1971).

Perrin, Jacques, *Winged Migration* (Paris: Galatée Films, 2001).

Perrin, Jacques, and Mongibeaux, Jean-François, *Winged Migration* (San Francisco: Chronicle Books; Paris: Editions du Seuil, 2003).

Popper, Karl R., *The Logic of Scientific Discovery* (London: Routledge, 2002 [1959]).

Schechner, Richard, *Between Theater and Anthropology* (Philadelphia: University of Pennsylvania Press, 1985).

Secord, James A., 'The crisis of nature', in *Cultures of natural history*, eds Nick Jardine, James A. Secord, and Emma C. Spary (Cambridge: Cambridge University Press, 1996), pp. 447–459.

Shapin, Steven, and Schaffer, Simon, *Leviathan and the Air-Pump* (Princeton: Princeton University Press, 1985).

Taylor, Rosie, 'BBC "fakes wildlife shots all the time": Veteran cameraman claims species 'smaller than rabbits' are filmed on custom-built sets', *Daily Mail*, 9 October 2013, accessed via <www.dailymail.co.uk/news/article-2450381/BBC-fakes-wildlife-shots-time-veteran-cameraman-claims.html>.

Weimerskirch, Henri & al., 'Energy saving in flight formation', *Nature*, 413 (2001), 697–698.

Willock, Colin, *The World of Survival* (London: André Deutsch, 1978).

CHAPTER 6

'Unmediated' Science Plays: Seeing What Sticks

Kirsten E. Shepherd-Barr

Abstract Kirsten E. Shepherd-Barr examines contemporary science theatre, with particular attention paid to interdisciplinary and experimental theatre emerging across Europe. These dramas reveal, Shepherd-Barr argues, that it is the process of working towards a piece of theatre rather than the finished product that is of greatest interest, both to audiences and to the theatre-makers themselves. Such theatrical performances invite extensive participation in meaning-making amongst all of those involved, including a range of scientific consultants. In conclusion, Shepherd-Barr reads these new dramas as extensively interdisciplinary and co-produced, leading to a new form of productively entangled epistemological experience.

The field of theatre and science is a vibrant one, flourishing and expanding in diverse directions. Over the past decade or so several full-length studies have been published as well as dozens of journal articles.[1] New directions in the field, such as intermediality, theatre and cognition, interspecies performance, and performance and medicine, are being explored in books and articles.[2] Scores of new plays and performances have also emerged, too numerous to list here, showing that the public appetite for science

K.E. Shepherd-Barr
University of Oxford, UK

© The Editor(s) (if applicable) and The Author(s) 2016
M. Willis (ed.), *Staging Science*,
DOI 10.1057/978-1-137-49994-3_6

105

on stage is not abating. But the field stretches further back; its pioneering works include Richard D. Altick's *The Shows of London* (1978), William Demastes's *Staging Consciousness* (2002), Jane R. Goodall's *Performance and Evolution in the Age of Darwin* (2002), and notable articles in *Interdisciplinary Science Reviews* (2002).[3] In short, theatre scholars and practitioners have been investigating the interconnections between science and performance for decades.

What is striking is that little of this theatre scholarship seems to register in the work of historians of science who are investigating the performance of science. Theatre scholars and historians of science may both be looking at the same performances and productions, but they are not in dialogue with each other. One of the aims of this collection is to address this problem, and this chapter sets out ways of doing this. History teaches valuable lessons in this regard. Both theatre and science underwent tremendous change, development, and transformation across the nineteenth and twentieth centuries, and both phenomena are characterized by an extraordinarily wide range of modes and practices. By the end of the nineteenth century, a certain narrowness sets in. Science splinters off into increasingly specialized areas. Theatre comes under the iron grip of realism and naturalism, so much so that most of modern drama afterwards is dedicated to countering this dominance. But for most of the nineteenth century, both science and theatre enjoyed a breadth and profusion of interests and styles that is surely unrivalled by other periods.[4] Theatre encompassed the 'legitimate' mainstream stage, with its comedies, dramas, well-made plays, farces, and melodramas, as well as 'illegitimate' forms such as street theatre, freak shows, commercially displayed peoples or ethnological exhibits, cabaret, vaudeville, and circus. Science was full of eminent researchers but it could also be done anywhere and by anyone: witness the popularity of amateur specimen-collecting and fossil-hunting. And, as recent work by Tiffany Watt-Smith and others has begun to show, including in this collection, the boundaries between scientific and theatrical activity were much more fluid and mutually constitutive than one might think. Scientists borrowed theatrical techniques in their experiments; theatre makers incorporated science into their plays and performances.

Much of the focus in studies by historians of science has been on the scientist as public figure, lecturing and demonstrating, rather than on theatrical performances. Undeniably, such scientific performances constitute a kind of theatre; scholars of performance studies would be the first to assert this. But historians of science studying these phenomena have not always brought to bear on them the full recognition of their status as performances and what that means within a performance studies and theatre history context. These science history studies have run parallel to the theatrical ones, only converging occasionally when the theatrical has encompassed historical context and the historical has encompassed performance elements; these have been welcome moments, but each would probably accuse the other side of not being sufficiently trained to deal adequately with its materials and perspectives.

One of the key issues for both sides is: how is the science engaged and 'packaged'? Elsewhere I have noted a tendency of successful 'science plays' like Michael Frayn's *Copenhagen* and Tom Stoppard's *Arcadia* to merge form and content, so that the play enacts the science.[5] My main concern here is to situate this idea of enactment within the larger framework of questions about the nature of interdisciplinarity. How does theatre's engagement with science fit with current models of, and debates about, the nature of interdisciplinary work? For years, historians of science, theatre historians and performance researchers have been exploring the diverse ways in which nineteenth-century science filtered into the many forms of performance that proliferated in the period. Altick, Goodall, and Tamsen Wolff on the one hand take theatre as their starting point; Bernard Lightman, Iwan Morus, Joe Kember et al. take science as theirs. While the findings have been very rich, the two scholarly grooves have been running in parallel. This volume is one of the first formal touchpoints between them and it shows the mutual benefits of paying attention to what we are each doing in our grooves and how much potential there is when we come together and cross paths. This is the vein I want to explore at the end of the chapter, when I turn to the larger framework of interdisciplinarity which, I argue, theatre that engages science is uniquely placed to address. In what follows, I will be focusing on contemporary productions that engage science, looking at how recent work has approached the fundamental question of how theatre engages most productively with science.

MEDIATING THE SCIENCE

As the vast array of styles of engaging science on stage shows, there is no one way of doing it. Having said that, biographical dramas about scientists have always enjoyed popularity and they constitute a substantial portion of science plays. A recent example, riding the wave of interest in the Darwin bicentennial in 2009, was Peter Parnell's play *Trumpery* (2009), which dramatized the years in Darwin's life from his receipt of Wallace's paper outlining natural selection—spurring Darwin to finish writing *On the Origin of Species* in a blaze of energy driven by fear of being 'trumped'—to the death of his beloved daughter Annie at age 10 which left Darwin devastated. The play charts these emotional highs and lows and depicts the colourful characters around Darwin: his wife Emma, his friends Hooker and Huxley, his enemy Richard Owen, and his gentle rival, Wallace.[6] There are entertaining discussions of evolution amongst the scientists, but the main interest of the play is in the biography and characterization; in theory, Parnell's Darwin might just as well be an eminent physicist or mathematician. Personal events trump scientific ideas. It is instructive to compare Parnell's approach with that of Timberlake Wertenbaker in her play *After Darwin* (1998), which imagines Darwin and FitzRoy on the *Beagle*. The play adopts a more complicated theatrical strategy by having two alternating time frames, the Darwin-FitzRoy period (1830s) and a contemporary setting (1990s). This structure makes the audience focus on how each period comments on the other, and this happens through the ideas as well as the doubling of actors in both time frames.[7] Of a different order is Frayn's *Copenhagen* (1998). While it uses biography in its depiction of Niels Bohr, his wife, Margrethe, and Werner Heisenberg and the mysterious circumstances of the latter's visit to Bohr in occupied Denmark in 1941, it stages the scientific ideas both men were passionately engaged with throughout their lives and which defined their relationship as much as political events did. Theatricality and science are deeply intertwined and interdependent in *Copenhagen*, in ways that elude *Trumpery* and other well-known 'science plays' like David Auburn's *Proof* (2001). As one critic has noted, '*Proof* isn't about mathematics; it's about family connections, genius and insanity, the quest for love and the search for truth.'[8]

In all of these examples, to greater or lesser degrees, science is mediated for the audience through biography. We may not be familiar with the scientific concepts but we recognize the people, and that familiarity helps us understand the ideas they are discussing. In short, it is rare for a play that depicts 'real' scientists to evade the pull of biography and focus on the science, especially if it is realistic or naturalistic in mode. Even plays like *Arcadia* (1993) that do not overtly portray 'real' people cannot entirely avoid the audience focusing on biography: for all its brilliant enactment of chaos theory and the second law of thermodynamics, the play relies on mediating its ideas through character, whether drawn from life or imagined.

Perhaps this is why Bertolt Brecht's play *Life of Galileo* (written in several versions between 1938 and 1946) is so exceptional and remains an important predecessor to the kinds of productions I will be discussing below. The play demonstrates the astronomy at its core even while it gives us a glimpse of Galileo the man and his circle of friends and enemies. Using epic theatre techniques, it anticipates events, it is episodic in structure, and it displaces expectations, for example shifting the emphasis from Galileo's famous recantation in court to how his students receive the news of it. Brecht refuses to allow the audience to sink into passive spectatorship by constantly exposing the workings of theatre and forcing us to analyze the events on stage for their causes rather than get swept up in emotional identification with the characters. He entertainingly enacts for us the heliocentric theory and performs experiments on stage, but he also raises much broader, profound questions about scientific method (for example advocating Popper's falsification theory that a scientific test must be able to be refuted, at least in principle) and whether the ends of science justify its means. Should scientific investigation be limited? Can it do more harm than good? And does science need heroes?

Galileo poses such questions successfully by finding a balance between biography and ideas through its deliberate refusal of realism. This kind of theatrical approach links Brecht to contemporary practitioners and theatre makers likewise suspicious of mediation through biography, such as the French director Jean-François Peyret and the Italian director Luca Ronconi. The difference is that they also want to avoid didacticism, certainly not something that troubled Brecht. 'We don't have to do night school' is Peyret's dismissive response to *Copenhagen,* questioning the kinds of epistemological models that plays like *Arcadia* and *Copenhagen* seem to present as they use biography, demonstration, metaphor, and other means of 'explaining' the science to the audience.[9] This resonates with the so-called turn to culture and away from biography that transformed the field of the history of science in the later 1980s and 1990s. Ronconi too insists that such crutches as biography and metaphor be removed: instead of explanation, just throw science and mathematics at the audience and 'see what sticks.'[10] These two directors have pioneered new kinds of 'science plays' that eliminate the playwright in favour of a collaboration between a director and a scientist, or a team of scientists. That such ventures seem to be occurring outside the UK is striking, and there may be several factors that account for it, though no definitive explanation has yet been given; these range from greater cultural receptivity and

predisposition towards experimental theatre (France) to financial invest-
ment by specific organizations such as Sigma Tau pharmaceutical company
(Italy). In Britain, the nearest equivalent to this kind of work has been the
ground-breaking and remarkable productions of Theatre de Complicite,
notably *A Disappearing Number* (2009) and *Mnemonic* (2002). I have
dealt extensively with some of these director-scientist collaborations elsewhere.[11] Here, I want to look at their most recent work, beginning with
Peyret's Ex Vivo/In Vitro (2011).

PROCESS OVER PRODUCT

Ex Vivo/In Vitro is an exploration of the new ethical issues and challenges thrown up by reproductive technology, as roles such as 'father'
and 'mother' and 'baby' are no longer stable identities. This topic has
received a lot of theatrical attention, most prominently by Carl Djerassi
and by Anna Furse.[12] It is analogous to the growing area of climate change
theatre in that both are utterly contemporary fields, introduced within the
past 50 years, that have only recently begun making it into the theatre,
and they are some of the few remaining realms 'where nature still rules the
game'.[13] In Ex Vivo/In Vitro Peyret and his collaborators, including neu-
roscientist Alain Prochiantz, take the audience through some of the new
and often alien ethical terrain being mapped out by advances in medicine
that are changing the way we conceive and the way we conceptualize time-
honoured familial categories.[14]

The first thing to notice about the play is that it has no playwright—at
least, not in the sense of a single author producing a finished text that
then gets acted and interpreted by a creative team. Peyret leaves the playwright out altogether, working instead with a team that includes scientists
and actors to develop a constantly evolving, and never completely finished, script. As with his previous experiments in science-theatre, Peyret
emphasizes process over product. Although you cannot buy a copy of 'the
play' to read or teach or share with your friends and colleagues, you can
read numerous versions, or 'partitions,' on the company's website, each
representing different stages of the script in its development. This emphasizes the play as a process of exploration, transparency, and interaction.
It also avoids the single viewpoint of any one person; it is theatre as the
result of dialogue and debate. This is entirely appropriate considering that
many of the subjects broached by the play itself are unresolved, 'open
questions' still up for debate in the wider culture, including gene therapy,

genetic manipulation, stem cell research, reproductive technology, and epigenetics. Since these questions relate to reproduction they also have greater relevance for the public compared to topics within specialized scientific fields like physics that seem far removed from everyday life.

This is only the latest in a long line of original productions by Peyret and his collaborators that engage science.[15] Ex Vivo/In Vitro bears the hallmarks of its predecessors: the use of recognizable real-life figures such as Darwin and Galileo, but a resistance to a biographical focus on them (such figures are pointedly absent from the stage, although present in titles of productions); the use of literary material, such as Kafka's short story 'An Address to the Academy' or Thoreau's *Walden*, that forms the basis for an exploration of related motifs, usually scientific; the use of striking stage images created through simple, original scenic design, such as giant movable and permeable membranes or a thousand ropes hanging down from the ceiling and covering the breadth and depth of the stage. These then become a necessary and integral part of the action rather than serving merely as scenery or backdrop to it.

Meanwhile, in Italy, the director Luca Ronconi, who died in 2015, worked in similar fashion to bring science to the stage, notably in *Infinities* (2000, 2002) and in *Biblioetica* (2006). Like Peyret, Ronconi sought ways around overt mediation of the science for the audience. He teamed up with the cosmologist John Barrow for *Infinities*, built on a bare-bones outline of several famous thought-experiments in mathematics that muse on the concept of infinity in different ways. Five such thought-experiments furnish the 'script': Hilbert's Hotel ('Hotel Infinity'), Living Forever, Infinite Replication Paradox, Kronecker vs. Cantor, and Time Travel. The play invites its audiences to place themselves in a range of scenarios and pose for themselves a series of questions. These are classic thought-experiments that will be familiar to most students of mathematics. For example, imagine you are the proprietor of a hotel with an infinite number of rooms. A new guest arrives and asks for a room; what do you do? Moving guests around an infinite number of rooms presents a mathematical headache. Or imagine that everyone lives forever. What happens to employment, life insurance, motivation? Doesn't endlessness engender apathy? Or imagine that you can travel through time. Does this really work in all possible ways? What if you could go back in time and kill your grandmother before she had children—what would happen to you? These are the kinds of abstract, 'what-if' scenarios presented by each of the thought-experiments Barrow provided, and they became the material of the play, in the form of scenes presented simultaneously in different spaces within the theatre on a given night. The scenarios can be shuffled; the order in which the audience encounters them does not matter. This gives the potential for an endless loop (an enactment of infinity). There is a great deal of mathematics in many of the scenarios, especially in Kronecker vs. Cantor and in Hilbert's Hotel, and it is simply thrown at the audience—it does not matter whether it sticks or not, just that it is there. A play of this nature, dealing with abstract issues, can afford to do this. Less straightforward and far more vexed is the question of whether a play like *Copenhagen* can afford to do the same, since so much more is at stake in a play that is not solely about ideas but concerns itself with a kind of science whose applications led to weapons of mass destruction and whose history is therefore inextricably and painfully bound up with real people, events, and politics.

Ronconi's next foray into science on stage, *Biblioetica*, was part of a larger theatrical project called *Domani* [*Tomorrow*], a devised project staged during the Turin Olympics in 2006. It consisted of five plays running concurrently in four different venues in Turin, addressing themes related to 'tomorrow': history, war, biotechnology, finance, and politics. *Biblioetica* was subtitled 'A Theatrical Reference Book' and the form it took bore this out literally. It was a living dictionary of ethical problems related to biomedical research, clinical problems, and case studies; a user's guide to negotiating the increasingly complicated moral terrain created by medical and scientific advances. Sample 'entries' or, as with *Infinities*, 'scenarios,' included: Informed Consent; Organ and Tissue Donation; Euthanasia; Tuskegee Study; Pain and Palliative Treatments; Consciousness; Embryo; Cloning; and, Bio-patents and Bio-piracy. But, as with all reference works, it was not intended to be seen in its entirety, only dipped into. The audience did not realize until well into the production that there were simultaneous scenes going on elsewhere in the theatre, perhaps even right on the other side of the wall of the room they were sitting in. What is omitted from reference works becomes as important as what is included; the emphasis is on the process of selection, which brings into focus the unsettling tension between seeming objectivity and human bias. The audience has to embrace both simultaneously, and the play seems to ask a basic but profound question: how is watching a factual dictionary entry enacted upon the stage—'cloning,' say, or 'organ transplantation'—different from looking that entry up in the dictionary? What happens when highly sensitive issues are conveyed by actors to a live audience in a seemingly objective fashion; does theatre by its very nature undermine any attempts at the kind of objectivity achieved by reference works?

THEATRICALITY AND INTERMEDIALITY

Ronconi's two productions involve several elements that intervene in the audience's experience of the production and shape its character. One is audience mobility: this is promenade-style theatre, the audience moving from one space to another, usually not sitting still for more than 15–20 minutes at a time. Another is the fact of omission: because some scenes run simultaneously, the audience will not necessarily see all scenes, and certainly not in the same order. In *Biblioetica*, the fact that the audience has to choose where it will go mimics the selectivity of consulting a reference work, seeing only certain entries. The play thus dramatizes the notion of choice involved in all ethical decisions and debates. The concept of choice operates on two levels: both epistemological (knowledge choice) and ethical. By 'staging' various biomedical dilemmas as if they were entries in a dictionary, the production allowed these two kinds of choice to converge.

For *Biblioetica*, there was a particular challenge: the production was in Italian, yet the audiences were highly international due to the Olympics. The solution was to provide individual 'palmare' [palm pilots] for audience members needing translations, so that they could select whichever language they preferred, for example, English, French, Italian, or German. This further underscored the theme of individual 'choice', and it meant that technology intruded on, and interrupted, the theatregoing experience; it transformed the nature of the event, instead of merely facilitating it. Audience members had to focus on written versions of the spoken text which they 'looked up' just as with a dictionary. Their attention was divided between spectating and reading. This also fractured the unity of the audience: though present in one room, they were not all experiencing the production in the same way linguistically. In short, the audience at a given performance was not unified by its spectatorship.

In addition, acting and design played a major part in this experience. Ronconi, much like Brecht before him, has his actors act *against* the text; there is an incongruity between what is being said and how it is said. In the scene discussing euthanasia, for example, there is anguish, because the parents are trying to decide whether to terminate the life of their daughter—a story taken from a real-life case with which audiences would perhaps be familiar. Yet actors would express themselves with detachment and dispassion, rather than sorrow or despair. Conversely, the actor discussing the science of cloning is histrionic and emotional although simply relaying facts. This defamiliarization is the hinge for much of Ronconi's theatre work. It extends to costuming: shades of grey and black dominate, with little colour in the clothes or indeed the surroundings of the actors; each room we enter is completely black, creating a stark and sterile atmosphere. The actors often wear half masks that cover the top of their faces leaving only their mouths moving, a further element of the defamiliarizing process. There are minimal props and furniture and the performance spaces have a clinical feel with their bare desks, operating tables, and trolleys, rather like a hospital. This makes the sudden appearance of the familiar, as in a life-like replica of Dolly the Sheep in the Cloning scene, startling, reassuring, and funny all at once, even as it makes you think about why you are reacting that way.

POSTDRAMATIC PERFORMANCE

These productions that I have been describing bear some of the characteristics of postdramatic theatre as outlined by Hans-Thies Lehmann in his ground-breaking 1995 book of that name. They revolve around simultaneity (here leading to omission). No single element of performance is foregrounded at the expense of another; this is what Lehmann calls the 'de-hierarchization of theatrical means.'[16] They lack a plot or *mythos*. They privilege presence over representation, manifestation over signification and, as we have seen, process over product. This is consistent with what Marco De Marinis calls 'the new theatrology.'[17] For all their spareness, these productions demonstrate an aesthetics of excess: a plethora of paratactical (or non-hierarchical) signs, a profusion of information, that an audience cannot possibly absorb and process in its entirety. These theatrical productions also depend on parataxis: images and texts juxtaposed without explicit connections. There is no single unifying perspective but rather fragmentation—in *Biblioetica*, for instance, the dictionary format merely serves as a point of contact between the two realms, not an instrument of harmonious reconciliation.

This last point is key, as it shows how these productions are not about 'bridging the gap' between the two cultures of science and theatre, but simply letting them co-exist on the stage. The aim is to avoid the unconscious hierarchy underpinning so many engagements between science and the arts, by which the science is privileged and the given art form subordinate; the science is the starting point and the theatre is enlisted to convey it. This hierarchy most often asserts itself when the goal is to transmit information to the audience. Freed of that goal, a production that engages with science for its own sake and not to teach but to suggest can risk the audience not learning anything at all in terms of 'real' science. Instead, the audience can experience whatever new knowledges come about from the interdisciplinary interaction on stage. In the final section of my discussion, I will map such productions onto the landscape of inter-, cross-, and transdisciplinary debates, seeing how these recent engagements between theatre and science might model a new and better way for radically disparate disciplines to meet.

INTERDISCIPLINARITY

The first question to ask is how these new kinds of science-theatre productions relate to broader concerns about interdisciplinarity—basic, fundamental questions about what we want, and what are we trying to achieve, when we put discrete fields together? Historians of science have criticized playwrights for their depictions of real-life scientists and events (for example, Paul Lawrence Rose's objections to Frayn's depiction of Werner Heisenberg in *Copenhagen* as too sympathetic). Scientists have criticized theatre for being scientifically inaccurate (for example, Ibsen's flawed understanding of the medical facts about syphilis in *Ghosts*). Theatre critics have criticized scientists writing plays that do not seem to work theatrically, even if the science is right (for example, criticisms of Carl Djerassi's plays for being static and preachy).[18] These various groups all argue over what makes a good 'science play' in the first place, and what should make the grade in the canon of this genre: can plays that were never staged make it in? Or do we have to have some measure of theatrical success (some kind of minimum run perhaps) to ensure canonicity? This, incidentally, would rule out some of the most important works, like Georg Büchner's *Woyzeck* (2000), while automatically including other works that may have enjoyed some transient success but lack enduring theatrical qualities and audience interest. New models such as 'science-in-theatre' as opposed to 'science theatre' have been proposed in an attempt to get around this problem, often reflecting an assumption that theatre is produced merely in the service of science. The intensity of the debates within this 'field' (or domain, or magisteria) of theatre and science testifies to how important and how vexed this issue is about the best interdisciplinary approach, and it suggests that this domain can act as a barometer of how interdisciplinarity works now.

To the general theatregoer, it might seem perplexing or mystifying that the use of science on stage should be so hotly contested. What is all the fuss about? Why should it matter *how* the science is being engaged, so long as the play is good? Of course, that is precisely the point. There is no general agreement on what a 'good' play is, what a 'good' night at the theatre might be. Audiences are not blocks or units, but complex and moving masses of individuals each with distinct responses to what they experience. Science plays might seem to require special preparation; David Auburn recounts how at *Proof*, only the mathematicians laughed at the maths jokes. But the foregoing examples of recent productions that have reacted against the 'seeing what sticks' model show that such insider's knowledge should not be required. The spectrum of science plays/science on stage is vast and encompasses widely divergent types of interaction so it requires a concomitantly capacious theory of interdisciplinarity that can take this array fully into account.

The problem is not that theatre and science are two unlike cultures joining together, because in fact they have much in common in terms of methodology. Theatre and science emphasize process and experiment, and rely on repetition and rehearsal of the same actions with ongoing permutations and adjustments to get the best possible outcome. In a very real sense, both domains are laboratories. Stuart Firestein has observed that the two domains also share a common interest in narrative: just as theatre makers tell a story, scientists too ask what *story* their data is trying to tell?[19] The problem now has more to do with stakeholders: who stands to gain from the interaction between theatre and science, particularly when funding agendas and 'public engagement' are involved? It is helpful, therefore, to turn to theories of interdisciplinarity that address such questions.

I want to begin by looking again at C.P. Snow's 'two cultures'. This phrase has become such common currency that it is easy to forget what Snow actually says in his lectures of 1959 published under that title. He talks of a 'clash of cultures', not a seamless merging of disciplines. He characterizes scientists and humanists/artists as 'two polar groups' separated by a 'gulf of mutual incomprehension'. He talks of scientists as 'self-impoverished,' with a limited 'imaginative understanding'. He chastises the English educational system for its 'fanatical belief' in specialization. He also says that these modes of thinking and patterns of education have 'crystallized'.[20]

The context for interdisciplinary work has changed a great deal since Snow's time; we now have, as Andrew Barry and Georgina Born see it, 'the evolving institutionalization of knowledge in the guise of science and research policy, research funding and evaluation, and the nature of the university'.[21] Helga Nowotny identifies a shift from 'science' to 'knowledge production' characterized by, among other elements, 'the growth of transdisciplinary research which, unlike interdisciplinary research, is not derived from pre-existing disciplines; [...] the displacement of a 'culture of autonomy of science' by a 'culture of accountability' [...] [and] a growing diversity of sites at which knowledge is produced'.[22] Recent thinking on interdisciplinarity proposes various new models based on this increasingly complex landscape. Barry and Born explain that their study *Interdisciplinarity* (2013) came about through their dissatisfaction with 'the teleological account of interdisciplinarity in much of the literature'.[23] By this they mean the assumption that disciplinarity is breaking down and that its 'progressive decline' is inevitable. This is a tenacious and popular idea. They wanted instead to 'get a sense of the multiplicity of interdisciplinary forms and their diverse histories, to interrogate the unity of interdisciplinarity', and to understand the concept 'less as a unity and more as a field of differences, a multiplicity'.[24] They also wanted to explore how ingrained it now is to understand interdisciplinarity in *instrumental* terms, 'terms that may inhibit rather than foster novelty'.[25] Is an 'antiinterdisciplinary' backlash inevitable? Barry and Born voice their 'pronounced scepticism about the value of interdisciplinarity "in general"'—a scepticism ameliorated only by the potential of interdisciplinary endeavour to be inventive.[26] Why bring two fields together unless you are aiming to generate something new that neither could generate on its own?

Whether or not science plays are models of such inventiveness depends on which angle you are approaching this question from. For Barry and Born, invention needs to be understood not as 'a moment in time, but as a process'.[27] For those in the scientific domain, the priority is usually the science: ensuring that the elements of theatre work together in conveying information accurately. Theatre becomes just another vehicle for such conveyance; a means by which scientific knowledge is faithfully communicated. For those in the theatrical domain—theatre practitioners and scholars—the emphasis will be on *how* (process) more than *what* (outcome or event). For them, the question is how an event engages an audience as a piece of performance, not whether the scientific ideas are accurately conveyed. Does the particular performance have artistic integrity? How does it use its chosen materials, space, actors, lighting, and costume? The focus is on the audience and its experience of the live event; the inventiveness is less in the show itself than in how it is received.

The liveness of theatre is the key ingredient here, distinct from film or other art forms in which the work of art is finished and then presented to a viewer whose response has no ability to affect the work itself. A general quality of theatrical events is that not everything can be under control or predictable; there may be a script, there may even be a director's ideal 'theoretical' production or an actor's ideal performance of a role, but there is no single, stable, unvarying, and predetermined outcome on the stage from night to night. It is not just that something new can emerge from the encounter between the work and the audience—as in the inherent inventiveness in the combination of science and theatre—but that that new entity paradoxically does not exist as a 'thing' but is entirely in the mind of the perceiver. Some may regard this instability and ephemerality as a drawback; others see it as the key, central quality that sets the genre apart from all others.[28]

A full consideration of the concepts of interdisciplinarity and transdisciplinarity, especially their historical development since the 1970s when these terms first emerged, lies beyond the scope of this chapter, but it is useful to point to a couple of aspects. Often, interdisciplinarity is linked to problem-solving; engaging with another discipline might lead to a solution to an intractable problem that a single discipline is not equipped to answer on its own. 'Multiple perspectives and collaborative work' can, theoretically, provide answers unavailable to the single discipline.[29] Meanwhile, transdisciplinarity 'is taken to involve a transgression against or transcendence of disciplinary norms.'[30] This idea of violation of existing rules or norms, challenging and transcending them, provides salient overlap with theatre. Transgression and challenging of norms is second nature to many theatre makers. In addition, theatre by its nature is already interdisciplinary in practice, combining many disparate elements and areas of expertise into a single project.

But what kind of project is produced by the interaction of theatre and science? Is it, as Snow would have it, a 'clash' of two sharply divergent cultures? Or is it an amicable merging, a meeting, a collaboration of equals? Do both retain their distinctiveness, or is one subsumed by the other? And if so, how consciously or unconsciously does this subordination happen and a hierarchy arrange itself? Above all, what are the assumptions on both the theatrical and the scientific side of the encounter about the other's domain? Barry and Born expose some of these assumptions: for example, in the art-science field, especially in the UK, 'funding has often been predicated on the notion that the arts are expected to provide a service to science, rendering it more popular or accessible to the lay public, or enhancing and publicising aesthetic aspects of scientific materials or imagery that might not otherwise be appreciated or known'.[31] Briefly, they identify several kinds of interdisciplinary activity at the present time, and they also taxonomize what they term the 'logics' of interdisciplinarity. There is a logic of accountability (e.g. the growth of public consultation, public understanding of science initiatives); a logic of innovation (e.g. anthropologists in IT and pharmaceutical industries); and a logic of ontology (e.g. experimental art-science, science and technology studies). It might seem that this last logic describes most accurately the theatre-science encounters I have been describing, since the experiment of bringing the two modes together is predicated on seeing what emerges rather than starting with a script already in place. The encounter therefore generates something entirely new. But there is another model that more aptly describes many of the theatrical encounters with science, because it leaves out the implicit obligation of theatre to be faithful in rendering the science: 'creative misprisions', a term used by Gillian Beer in her book *Open Fields: Science in Cultural Encounter* and alluding to Harold Bloom's usage of the term in his seminal study *The Anxiety of Influence*. She gives as one of her examples Tony Harrison's play *Square Rounds*, a play that explores the work and legacy of Fritz Haber, whose discovery of fertilizer both helped and harmed (benefiting agriculture but also being used during World War I to create mustard gas). She shows that the fundamental quality of *Square Rounds* is that it doesn't simply translate the science at its core (in this case, chemistry) but transforms it into something new.

This kind of transformation happens when theatre meets science not only because of the multiple elements at work in any theatrical event but also, crucially, the audience's participation in the construction of meaning. Productions like *Infinities* and Ex Vivo/In Vitro do not teach the science (one-way, top-down engagement) but instead risk, or perhaps even encourage, creative misprision by using suggestion—through visual, textual, scenic, and aural means—and leaving it to the audience to make connections, fill in blanks, deduce, and interpret the science. The text becomes just another component in the theatrical event, not the single most important element to which all others are subservient. In fact, some of the most memorable moments in these productions are non-textual. In *Infinities*, for example, the languid, unhurried atmosphere of the 'Living Forever' paradox scenario, with its impossibly aged bodies, or the eerie 'Library of Babel' scene with its simple trick of mirrors and identically clad actors, convey a sense of endlessness and the idea of infinite replication. In *Ex Vivo/In Vitro* a thousand thick ropes hang from the ceiling like a forest of umbilical cords, and the actors interact imaginatively with them to denote everything from DNA, chromosomes, childbirth, filiation, parenthood, and the concept of nurture. All these examples show an engagement with science that allows the audience to 'make' the knowledge itself through the power of suggestion; a different epistemological experience altogether from 'night school.'

NOTES

1. Kirsten E. Shepherd-Barr, *Theatre and Evolution from Ibsen to Beckett* (New York: Columbia University Press, 2015) and *Science on Stage: From Doctor Faustus to Copenhagen* (Princeton: Princeton University Press, 2006, 2012); Sue-Ellen Case, *Performing Science and the Virtual* (New York: Routledge, 2006), Tamsen Wolff, *Mendel's Theatre: Heredity, Eugenics, and Early Twentieth-Century American Drama* (New York: Palgrave Macmillan, 2009), Eva-Sabine Zehelein, *Science: Dramatic: Science Plays in America and Great Britain, 1990–2007* (Heidelberg: Universitätsverlag Winter, 2009), and Liliane Campos, *Sciences en scène* (Rennes: PUR, 2012).

2. See for example two recent special issues of *Interdisciplinary Science Reviews* on 'New Directions in Theatre and Science', co-edited by Kirsten E. Shepherd-Barr and Carina Bartleet, 38.4 (2013) and 39.3 (2014); a special issue of *Theatre Journal* edited by Bruce McConachie on 'Performance and Cognition', 59.4 (2007); and a forthcoming book, *Performance and the Medical Body*, eds Alex Mermikides and Gianna Bouchard (Methuen Bloomsbury, 2016). 'Intermediality' refers to the use of different media within a single work of art.

3. Richard D. Altick, *The Shows of London* (Cambridge: Belknap Press, 1978), William Demastes, *Staging Consciousness: Theater and the Materialization of Mind* (Ann Arbor: University of Michigan Press, 2002), Jane R. Goodall, *Performance and Evolution in the Age of Darwin: Out of the Natural Order* (London: Routledge, 2002), and *Interdisciplinary Science Reviews* 27.3 (2002), 38.4 (2013), and 39.3 (2014).

4. *Evolution and Victorian Culture* eds Bernard Lightman and Bennett Zon (Cambridge: Cambridge University Press, 2014); *Popular Exhibitions,*

Science and Showmanship, 1840–1910 eds Joe Kember, John Plunkett, and Jill A. Sullivan (London: Pickering and Chatto, 2012); Bernard Lightman, *Victorian Popularizers of Science: Designing Nature for New Audiences* (Chicago: University of Chicago Press, 2007) ; and Goodall, *Performance and Evolution.*

5. Shepherd-Barr, *Science on Stage*, Chaps. 4 and 6.
6. Shepherd-Barr, 'Darwin on Stage: Evolutionary Theory in the Theatre', *Interdisciplinary Science Reviews*, 33.2 (2008), 107–115.
7. Shepherd-Barr, *Science on Stage*, pp. 114–122.
8. Karen C. Blansfield, 'Atom and Eve: The Mating of Science and Humanism', *South Atlantic Review*, 68.4 (2003), 1–16.
9. Jean-François Peyret, interview with the author, Cambridge, England (March 2004), trans. Lisbeth Shepherd, quoted in Shepherd-Barr, *Science on Stage*, p. 202.
10. Luca Ronconi, interview with the author, Teatro Piccolo, Milan (May 2003), trans. Pino Donghi.
11. Shepherd-Barr, *Science on Stage*, pp. 199–218, and 'Des 'Liens significatifs': Luca Ronconi et les scientifiques' ['"Meaningful Joinings": Luca Ronconi and the Scientists'], *Alternatives théâtrales*, 102–103 (2009), 28–33; and Shepherd-Barr and Liliane Campos, 'Open Dialogues between Science and Theatre: *Biblioetica, Le Cas de Sophie K.,* and the Postdramatic Science Play', *Interdisciplinary Science Reviews* 31.3 (2006), 245–253.
12. Carl Djerassi, *An Immaculate Misconception* (London: Imperial College Press, 2001) and Anna Furse, *Yerma's Eggs* (2003), accessed via http://www.gold.ac.uk/theatre-performance/research/practice-as-research/drama/research/furse-eggs/; and see Furse, 'Hospital Drama: Visual Theatres of the Medical Rendezvous from Asylum to Hospital with Reference to Specific Works by Anna Furse', *Interdisciplinary Science Reviews* 39.3 (2014), 238–257.
13. From Peyret's introduction to the play on his company's website, accessed via http://theatrefeuilleton2.net/spectacles/ex-vivo-in-vitro/
14. Detailed information on the play and its production history can be found on the website of Theatre Feuilleton 2, accessed via http://theatrefeuilleton2.net/spectacles/ex-vivo-in-vitro/
15. Other productions are listed on the company website, accessed via http://theatrefeuilleton2.net/spectacles/
16. Hans-Thies Lehmann, *Postdramatic Theatre*, trans. Karen Jürs-Munby (London: Routledge, 2006), p. 86.
17. Marco De Marinis, trans. Marie Pecorari, 'New Theatrology and Performance Studies', *TDR: The Drama Review*, 55.4 (2011), 64–74.
18. For a comprehensive example of all three of these trends, see *Interdisciplinary Science Reviews*, 27.3 (2002).

19. Stuart Firestein, *Ignorance: How it Drives Science* (New York: Oxford University Press, 2012).
20. C.P. Snow, *The Two Cultures: And a Second Look* (Cambridge: Cambridge University Press, 1963), pp. 16, 25, 71.
21. *Interdisciplinarity: Reconfigurations of the Social and Natural Sciences*, eds Andrew Barry and Georgina Born (London: Routledge, 2013), p. 1.
22. Helga Nowotny, Peter Scott and Michael Gibbons, *Re-Thinking Science: Knowledge and the Public in an Age of Uncertainty* (2001), cited in Barry and Born, pp. 1–2.
23. Barry and Born, eds., *Interdisciplinarity*, p. 4.
24. Barry and Born, eds., *Interdisciplinarity*, p. 5.
25. Barry and Born, eds., *Interdisciplinarity*, p. 5.
26. Barry and Born, eds., *Interdisciplinarity*, p. 5.
27. Barry and Born, eds., *Interdisciplinarity*, p. 6.
28. Jure Gantar, 'Catching the Wind in a Net: The Shortcomings of Existing Methods for the Analysis of Performance', *Modern Drama*, 39.4 (1996), 537–546, and Emma Smith, '"Freezing the Snowman": (How) Can We Do Performance Criticism?', in *How to Do Things with Shakespeare: New Approaches, New Essays*, ed. Laurie Maguire (Blackwell, 2008), pp. 280–297.
29. M. Strathern, "Social Property: An Interdisciplinary Experiment" (2004), quoted in Barry and Born, eds., *Interdisciplinarity*, p. 8.
30. Barry and Born, eds., *Interdisciplinarity*, p. 9.
31. Barry and Born, eds., *Interdisciplinarity*, p. 11.

BIBLIOGRAPHY

Altick, Richard D., *The Shows of London* (Cambridge: Belknap Press, 1978).
Barry, Andrew, and Georgina Born, eds, *Interdisciplinarity: Reconfigurations of the Social and Natural Sciences* (London: Routledge, 2013).
Blansfield, Karen C., 'Atom and Eve: The Mating of Science and Humanism', *South Atlantic Review*, 68.4 (2003), 1–16.
Campos, Liliane, *Sciences en scène* (Rennes: PUR, 2012).
Campos, Liliane, and Kirsten Shepherd-Barr, 'Open Dialogues between Science and Theatre: *Biblioetica, Le Cas de Sophie K.*, and the Postdramatic Science Play', *Interdisciplinary Science Reviews*, 31.3 (2006), 245–53.
Case, Sue-Ellen, *Performing Science and the Virtual* (New York: Routledge, 2006).
De Marinis, Marco, translated Marie Pecorari, 'New Theatrology and Performance Studies', *TDR: The Drama Review*, 55.4 (2011), 64–74.
Demastes, William, *Staging Consciousness: Theater and the Materialization of Mind* (Ann Arbor: University of Michigan Press, 2002).

Djerassi, Carl, *An Immaculate Misconception* (London: Imperial College Press, 2000).

Firestein, Stuart, *Ignorance: How it Drives Science* (New York: Oxford University Press, 2012).

Furse, Anna, 'Hospital Drama: Visual Theatres of the Medical Rendezvous from Asylum to Hospital with Reference to Specific Works by Anna Furse', *Interdisciplinary Science Reviews*, 39.3 (2014), 238–57.

Gantar, Jure, 'Catching the Wind in a Net: The Shortcomings of Existing Methods for the Analysis of Performance', *Modern Drama*, 39.4 (1996), 537–46.

Goodall, Jane R., *Performance and Evolution in the Age of Darwin: Out of the Natural Order* (London: Routledge, 2002).

Kember, Joseph, John Plunkett, and Jill A. Sullivan, eds, *Popular Exhibitions, Science and Showmanship, 1840–1910* (London: Pickering and Chatto, 2012).

Lehmann, Hans-Thies, *Postdramatic Theatre*. Trans. Karen Jürs-Munby (London: Routledge, 2006).

Lightman, Bernard, *Victorian Popularizers of Science: Designing Nature for New Audiences* (Chicago: University of Chicago Press, 2007).

Lightman, Bernard, and Bennett Zon, eds, *Evolution and Victorian Culture* (Cambridge: Cambridge University Press, 2014).

Nowotny, Helga, Peter Scott and Michael Gibbons, *Re-Thinking Science: Knowledge and the Public in an Age of Uncertainty* (Cambridge: Polity, 2001).

Peyret, Jean-François, accessed via <http://theatrefeuilleton2.net/spectacles/ex-vivo-in-vitro/>.

Shepherd-Barr, Kirsten, *Science on Stage: From Doctor Faustus to Copenhagen* (Princeton: Princeton University Press, 2006, 2012).

Shepherd-Barr, Kirsten, 'Darwin on Stage: Evolutionary Theory in the Theatre', *Interdisciplinary Science Reviews*, 33.2 (2008), 107–15.

Shepherd-Barr, Kirsten, *Theatre and Evolution from Ibsen to Beckett* (New York: Columbia University Press, 2015).

Shepherd-Barr, Kirsten, 'Des 'Liens significatifs': Luca Ronconi et les Scientifiques' ['"Meaningful Joinings": Luca Ronconi and the Scientists'], *Alternatives théâtrales*, 102–3 (2009), 28–33.

Smith, Emma, '"Freezing the Snowman": (How) Can We Do Performance Criticism?', in *How to Do Things with Shakespeare: New Approaches, New Essays*, ed. Laurie Maguire (Malden, MA: Blackwell, 2008), pp. 280–97.

Snow, C.P., *The Two Cultures: And a Second Look* (Cambridge: Cambridge University Press, 1963).

Wolff, Tamsen, *Mendel's Theatre: Heredity, Eugenics, and Early Twentieth-Century American Drama* (New York: Palgrave Macmillan, 2009).

Zehelein, Eva-Sabine, *Science: Dramatic: Science Plays in America and Great Britain, 1990–2007* (Heidelberg: Universitätsverlag Winter, 2009).

Afterword

Bernard Lightman

Abstract In this short afterword, the leading historian of science Bernard Lightman reflects upon the importance and centrality of scientific performance to science and its histories.

The physicist John Tyndall (1820–1893) was one of most well-known and admired science lecturers of the second half of the nineteenth century. Indeed, his style and methods remain influential in contemporary scientific performances. From 1853 to 1887 he lectured to the wealthy and fashionable audiences at the Royal Institution on Albemarle Street in London. Tyndall's reputation as one of the foremost lecturers of the day was based on his ability to explain complex scientific theories through the staging of spectacular experimental demonstrations. His lectures therefore provide excellent examples of how even elite scientific figures adopted the role of performer when they communicated to a Victorian audience. During one Friday evening discourse, he aimed to recreate an alpine rainbow in the lecture hall. After condensing steam from a high-pressure boiler till the air was full of minute droplets, he then set up an arc light in a camera with a condensing lens in front of it. Standing in the midst of his artificially illuminated cloud, and dressed in protective rain gear, Tyndall

B. Lightman
York University

© The Editor(s) (if applicable) and The Author(s) 2016

M. Willis (ed.), *Staging Science*,

DOI 10.1057/978-1-137-49994-3

described to the audience all the colours that were produced.[1] While hundreds of thousands of Londoners saw an astonishing display of the Ascent of Mont Blanc in the mid-1850s at the Egyptian Hall, Tyndall created a demonstration of Alpine glacier movement and the ice flowers that were produced as a result. Tyndall later described the experiment in an article: 'A slab of ice was prepared and placed in front of an electric lamp: by a lens placed in front of the slab, the latter was projected upon a screen; on sending a beam from the lamp through the ice, the formation of [ice] flowers was rendered visible to the audience'.[2] Tyndall's demonstrations featuring visual displays were so effective that the physicist James Clerk Maxwell confessed to a friend that he was preparing for a lecture on colour by 'Tyndalising my imagination'.[3]

Ensuring that these demonstrations worked was no easy task. Countless hours of preparation were required so that Tyndall's performances appeared to be effortless. He wrote out his discourses in full, producing a final version only after going through several rough drafts. Tyndall also spent time mastering the apparatus to be used in order to present the most lucid and visible demonstrations. He experimented for hours with different combinations of apparatus until he was certain which one would guarantee the best performance. He practised his experiments repeatedly, timed them, and wrote them out in the order in which they were to be performed so that his assistants were fully informed. Tyndall left nothing to chance.[4] But he was also willing to use chance accidents to his advantage to heighten the theatricality of his lectures. Once, he accidentally knocked a flask off a desk while preparing for a lecture, but quickly jumped over the desk and caught the flask before it hit the ground. He then rehearsed the 'accident' and included it in his lecture. He knew that his audience would be impressed when they saw him leaping over the lecture table to catch the falling flask.[5] Tyndall's painstaking preparations and his emphasis on rehearsal were similar to the amount of effort that normally goes into the production of a theatrical play.

Tyndall was sometimes criticized for his theatrical lecturing style. His enemies tried to undermine his reputation as a scientist by claiming that he spent too much time devising spectacular experiments instead of doing serious research. But even his friends could be critical. His closest friend, the mathematician Thomas Archer Hirst, noted in his private journal that Tyndall tried too hard to entertain the women in his audience by repeating spectacular experiments. Tyndall's strong liking for the theatrical, and not just in his lectures, puzzled Hirst. It seemed like a contradiction in Tyndall's character. On the one hand, Tyndall was fascinated by the most complex philosophical figures, such as Kant and Fichte. On the other, he enjoyed the theatre. One night after attending a play with Tyndall, Hirst wrote in his journal: 'I sometimes wonder that Tyndall has preserved his liking for the theatre amidst his high occupations. As I stood by him in the pit watching him laugh at the nonsense before us [...] and reflected that he had just come from writing his philosophical chapter on the relation between Thought and Physics. The contrast appeared remarkable'.[6]

The theatricality of Tyndall's lectures is emblematic of the main theme of this collection, the two-way traffic between modern science and the stage. In the last three decades scholars have examined a similar dialectic at work in the intricate intersections between science and culture, whether it be the musical, literary, photographic, artistic, or the architectural dimensions of science. The deep relationship between science and all forms of culture is an endlessly fascinating topic of study. But the linkage between science and performativity is not as well studied as many of the others, and, as these chapters have proven, it is a particularly rich subject to explore. Exploring this specific link has taken us to startlingly diverse spaces, from theatres, movie houses, and the haunts of London tourists, to the lecture hall, the wilderness, and the laboratory. Along the way we have also encountered a wide variety of people, including physicists, inventors, travellers, psychologists, popularizers, movie-makers, animal trainers, directors, playwrights, and actors. But perhaps most impressive of all is the cross-disciplinary nature of the chapters required to track this diverse list of people and places. The contributors have had to make use of different disciplinary perspectives in order to understand the performative feature of science and the scientific quality of theatre.

The previous research of theatre scholars and historians of science has not been as interdisciplinary. Shepherd-Barr describes it as running in two parallel, but separate, grooves, although at times scholars are looking at the same performances. The collection addresses this problem in two ways, by exploring what science and theatre share in common, and by suggesting why this commonality exists. They share a similar methodology, since they both depend on process and experiment, which require repetition and rehearsal to guarantee optimal results. In this sense, actors and scientists both work in laboratory environments. They have a mutual interest in narrative, as plays tell stories while scientists try to determine what tale their data tells them about the natural world. Both the scientist and the actor present themselves publicly through performance in order to tell that story.

The common characteristics of theatre and science exist since historically they have developed together, sometimes shaping each other in the process. The chapters have highlighted periods when the boundary between science and theatre was especially porous. In the second half of the nineteenth century, for example, sometimes referred to by historians as the age of the worship of science, science was ubiquitous—as Willis shows through his discussion of the large number of scientific sites included in London guidebooks. At this same time, as Morus observes, the use of spectacular demonstrations was central to Victorian physics. For Shepherd-Barr, contemporary productions, such as Peyret's Ex Vivo/In Vitro (2011) and Ronconi's *Infinities* (2000, 2002) and *Biblioetica* (2006), provide a clear window onto how theatre engages productively with science. It is here that we can see what is hidden from view by narrow disciplinary approaches, as well as by the notion that science is sui generis and completely separate from the rest of human culture.

By adopting an interdisciplinary approach the contributors provide us with a series of insights on the important role of performativity in science and theatre. Shepherd-Barr explores what happens when actors play scientists in modern, experimental theatrical productions. Of course we expect actors to perform, but several chapters deal with the performativity of scientists. Morus helps us to understand that Tyndall should be considered as part of a performative tradition in nineteenth-century physics that included Warren de la Rue, Cromwell Varley, William Crookes, and J.J. Thomson. Watt Smith demonstrates that developmental psychologists in the UK's first major psychological laboratory at University College, London, were required to perform in experiments on the behaviour of children. But it is not just scientists who perform. So do animals and institutions. Gouyon explores the use of birds as actors in wildlife film-making while Willis reveals how scientific sites, such as the Crystal Palace, perform as monuments to empire in Victorian travel guides to London.

Another insight that runs through the collection concerns the frequent use of spectacle in performance. Science and theatre often rely on spectacle to offer a visually appealing performance that engages the audience. In some cases, visitors were treated to a spectacle before even entering a scientific institution. The buildings themselves could be designed to impress the senses. As Willis reminds us, scientific sites, such as the Crystal Palace and the Woolwich Arsenal, were treated in London guidebooks as being representative of the gigantic scale of London. Not only were they massive structures, they were also architectural innovations. The Panopticon in Leicester Square, which opened in 1854, was constructed to resemble an enchanted Moorish palace, while the Colosseum was built in imitation of the Pantheon at Rome.

Once inside the building the experiments, demonstrations, and displays bombard the senses with visual stimuli. By focusing on Gassiot's cascade, one of the most dazzling and extensively used electrical demonstrations of the second half of the nineteenth century, Morus not only sheds light on the fluid boundary between science and theatre, he also links it to the larger context of a Victorian visual culture of spectacle and spectacular performance. Spectacular experimental displays like Gassiot's cascade were featured throughout the nineteenth century at many of the important scientific sites of London, such as the Adelaide Gallery, the Royal Polytechnic Institution, Wyld's Globe, the Panopticon, and the Colosseum. It is important to note that generating spectacle was considered to be one and the same as generating knowledge. But spectacular experiments were also a necessity when competing with entertainments in London such as Burford's Panorama in Leicester Square. The theatre was also home to spectacle. Panoramic and dioramic spectacle had begun to be widely used in dramatic productions in London in the 1820s, and later, in the 1850s, Victorians were attracted by the Shakespearean revivals by Charles Kean and Samuel Phelps, which offered magnificent spectacles.[7]

Successful scientific and theatrical performances involve expertise, whether it is the manipulation of test tubes, the use of props or the production of large-scale special effects in order to generate a spectacle. Gouyon discusses the expertise needed to stage a wildlife film. The professionalization of natural history film-making requires the reconstruction of natural phenomena. In the case of a film about bird migration the complex staging involves the transportation, care, training, and management of unruly fowls. Watt Smith explores how developmental psychologists mastered the ability to work with props and costumes, especially hats, in order to study the laughter of unruly children. The degree of expertise needed in both of these cases was no different from that required by nineteenth-century scientific performers who worked with temperamental instruments, such as gigantic electrical machines.

Another insight that runs throughout a number of the chapters has to do with the link between expertise and the production of realistic effects. Expertise is needed in both science and on stage in order to persuade the audience that what has been created artificially is actually real. Gouyon points to the complex techniques used by wildlife film-makers to provide a simulation of nature on the screen. A naive commitment to observational realism, so important before 1970, gave way to a new, more self-reflexive attitude towards depicting nature that acknowledged the work that went into recreating nature. Similarly, the whole purpose of panoramas, phantasmagoria, and scientific spectacle is to deceive the senses by manipulating the eye. Sometimes the fallibility of vision was treated as a part of the spectacle itself. John Henry Pepper, for example, after stunning his audience with a brilliant illusion, explained how scientific knowledge was used to produce it. A similar creation of artful illusions of reality is a feature of theatre practice.

A final insight gained by adopting an interdisciplinary perspective has to do with how identity is created in the performances to be found in the laboratory, in the lecture hall, and on the stage. Shepherd-Barr discusses how contemporary plays use the moral issues raised by modern science to force us to define who we are by our choices. Watt Smith investigates how the power of hats to transform the identity of the wearer figured into the techniques that Sully and his team developed to make childish laughter accessible to scientific scrutiny. To Morus, the making of a public scientific self was integrally linked to how epistemic authority was exhibited in performance. But it is not just the individual's identity that is affected. For Willis, travel guides invite the reader to act out their role as tourist and to create, with the help of scientific sites, new forms of urban and national identity. In short, it is through performance that identity is created and maintained.

I began with an account of one well-known nineteenth-century scientific lecturer, John Tyndall. I end with an anecdote about another. On April 7, 1821, a notice appeared in the *Morning Chronicle* in the 'Mirror of Fashion' section. A new astronomical lecture was to be presented at the English Opera House of the Theatre Royal. 'The whole of the Machinery necessary for a popular illustration of the sublime phenomena of the Heavens', the *Morning Chronicle* announced, 'has been constructed on a scale of unprecedented splendour, magnificence, and expense'. The new apparatus would display spectacular transparent scenes representing the planets, comets, and eclipses. The lecturer was George Bartley.[8] Bartley's lectures, given during Lent from 1821 to 1828, were well received by his contemporaries. London guidebooks considered his lectures to be superior to previous astronomical discourses. At least one significant scientific figure was also impressed. William Kitchener, Fellow of the Royal Society, optical instrument enthusiast, and writer, recommended the lectures and claimed that Bartley's fame as an astronomical lecturer was well deserved.[9]

Although Bartley was considered to be a successful scientific lecturer, he had little formal training in science. He was in fact a professional actor. Before he became involved in astronomical lecturing he had appeared on the London stage and in the provinces. His London début as Orlando in *As You Like It* took place in 1802. He later unsuccessfully managed the Glasgow theatre, and then established himself as a comedic actor in Manchester, Liverpool, and other towns. After marrying Sarah Smith in 1814, a tragic actress whose reputation eclipsed him, the couple made a triumphant trip to America in 1818. When he returned to England he accepted roles in plays at Covent Garden and the Lyceum, and then become involved in astronomical lecturing.[10] Bartley drew upon his skills as a showman cultivated from his theatrical profession to succeed as science lecturer.[11] But the ease with which he was able to move from the theatrical stage to the scientific lecturing platform is a reminder of the central theme of this collection: performativity connects the world of drama with the world of knowledge.

NOTES

1. John Tyndall, *New Fragments* (New York: D. Appleton, 1892), pp. 214–215 (p. 223); Simon Schaffer, 'Transport Phenomena: Space and Visibility in Victorian Physics', *Early Popular Visual Culture*, 10.1 (2012), 71–91 (p. 78).

2. John Tyndall, 'On Some Physical Properties of Ice', *Philosophical Transactions of the Royal Society*, 148 (1858), 213; Schaffer, 'Transport Phenomena', 79.

3. Lewis Campbell and William Garnett, *The Life of James Clerk Maxwell* (London: Macmillan, 1882), p. 379.

4. Jill Howard, '"Physics and Fashion": John Tyndall and His Audiences in Mid-Victorian Britain', *Studies in History and Philosophy of Science*, 35 (2004), 729–758 (p. 734).

5. N. D. McMillan and J. Meehan, *John Tyndall: 'Xemplar of Scientific and Technological Education* ed. Pauric Hogan (Dublin: ETA Publications, 1980), p. 49.

6. Hirst Journal, 19 February 1860, in *Natural Knowledge in Social Context: The Journals of Thomas Archer Hirst FRS* eds W. H. Brock and R. M. MacLeod (London: Mansell, 1980); Howard, 'Physics and Fashion', 746.

7. David Mayer III, *Harlequin in His Element: The English Pantomime, 1806–1836* (Cambridge, MA: Harvard University Press, 1969), pp. 69–70; Martin Meisel, *Realizations: Narrative, Pictorial, and Theatrical Arts in Nineteenth-Century England* (Princeton: Princeton University Press, 1983), p. 33, p. 62, pp. 380–384; Emily Allen, *Theatre Figures: The Production of the Nineteenth-Century British Novel* (Columbus: Ohio State University Press 2003), p. 5, pp. 20–21.

8. Anon., 'The Mirror of Fashion. Theatre Royal, English Opera House', *Morning Chronicle*, 16210, 7 April 1821, p. 3.
9. Hsiang-Fu Huang, *Commercial and Sublime: Popular Astronomy Lectures in Nineteenth Century Britain*, unpublished Ph.D. thesis, University College London, 2015, pp. 117–119.
10. Joseph Knight, 'Bartley, George (1782?–1858)', rev. Katharine Cockin, *Oxford Dictionary of National Biography* (Oxford: Oxford University Press, 2004), accessed via http://www.oxforddnb.com.ezproxy.library.yorku.ca/view/article/1589
11. Hsiang-Fu Huang, *Commercial and Sublime*, p. 118.

Bibliography

Allen, Emily, *Theatre Figures: The Production of the Nineteenth-Century British Novel* (Columbus: Ohio State University Press 2003).

Anon., 'The Mirror of Fashion. Theatre Royal, English Opera House', *Morning Chronicle*, 16210, 7 April 1821, p. 3.

Brock, W. H. and R. M. MacLeod, eds *Natural Knowledge in Social Context: The Journals of Thomas Archer Hirst FRS* (London: Mansell, 1980).

Campbell, Lewis and William Garnett, *The Life of James Clerk Maxwell* (London: Macmillan, 1882).

Howard, Jill, '"Physics and Fashion": John Tyndall and His Audiences in Mid-Victorian Britain', *Studies in History and Philosophy of Science*, 35 (2004), 729–758.

Huang, Hsiang-Fu, *Commercial and Sublime: Popular Astronomy Lectures in Nineteenth Century Britain* (Unpublished Ph.D. thesis, University College London, 2015).

Knight, Joseph, 'Bartley, George (1782?–1858)', rev. Katharine Cockin, *Oxford Dictionary of National Biography* (Oxford: Oxford University Press, 2004).

McMillan, N. D. and J. Meehan, *John Tyndall: 'Xemplar of Scientific and Technological Education*, Ed. Pauric Hogan (Dublin: ETA Publications, 1980).

Mayer, David III, *Harlequin in His Element: The English Pantomine, 1806–1836* (Cambridge: Harvard University Press, 1969).

Meisel, Martin, *Realizations: Narrative, Pictorial, and Theatrical Arts in Nineteenth-Century England* (Princeton, NJ: Princeton University Press, 1983).

Schaffer, Simon, 'Transport Phenomena: Space and Visibility in Victorian Physics', *Early Popular Visual Culture*, 10.1 (2012), 71–91.

Tyndall, John, *New Fragments* (New York: D. Appleton, 1892).

Tyndall, John, 'On Some Physical Properties of Ice', *Philosophical Transactions of the Royal Society*, 148 (1858).

INDEX

A

accountability, 117, 119
actor-network theory, 61
Addyman, C., 59–62, 65
Adelaide Gallery, 20, 22–3, 129
'affectations of the soul,' concept of, 63
After Darwin, 108
Allan, D., 84
Alpine glacier, demonstration of, 126
Altick, R.D., 106, 107
'An Address to the Academy,' 111
An Essay on Laughter
  'baffling spirit,' 64
  descending incongruity, 63–4
  fullness and complexity, 66
  hat-based comic scenarios, 67
  innermost feelings, 63
  mechanical objectivity, virtue of,
    65, 76
  'new child-study,' 65
  order and law, 65
  plurality of causes, 64
  primitive societies, 66–7
  professional discipline, psychology, 62
  psychology's 'expert knowledge,' 62

relief theory, 63
'sudden glory,' 63
theatrical performance,
  experiments, 66
untrustworthy emotion, 64
The Anxiety of Influence, 119
Arcadia, 107
artificiality
  captive or tame animals, use of, 95
  cognitive legitimisation, 94
  Making Wildlife Movies, 94
  MODs, 96
  unarmed hunters, natural history
    film-makers, 95–6
Ascent of Mont Blanc, 126
As You Like It, 131
aurora borealis, 17
Austin, J.L., 61
avian actors, 88

B

Baedeker's *London and its Environs*, 42
  arsenal's scale, 46
  South Kensington museums, 49

© The Editor(s) (if applicable) and The Author(s) 2016   133
M. Willis (ed.), *Staging Science*,
DOI 10.1057/978-1-137-49994-3

Bain, A., 63, 76
Barrow, J., 111–12
Bartlett, D., 85, 92–4
Bartlett, J., 85, 92–4
Bartley, G., 130–1
BBC Natural History Unit
    (NHU), 95
Beer, G., 119
behind-the-scene sequences, 89
Benjamin, W., 50–1
Biblioetica, 111–14, 128
Billig, M., 74
Black's Guide to Edinburgh, 41
*Black's Guide to London and its
    Environs*
    hierarchy of tourist sites, 43
    modernity, 48
    Zoological Gardens, 48
Blanchard, E., 40, 43–51
Blue-Chip documentaries, 86–7
The Blue Planet, 84
Bohr, N., 108
Bose, G., 17
Boyle, R., 85
Bradshaw, G., 36–7, 39, 51
*Bradshaw's Descriptive Railway
    Handbook*, 35–7
*Bradshaw's Handbook of Great
    Britain*, 48
*Bradshaw's Illustrated Handbook to
    London*, 42
    city's global significance, 43
    Herschel's claim, 43
    industrial events, 39
    new Westminster Bridge, 49–50
    railway timetables, 39
    Royal Institution, 45
    scientific sites, 45
Brecht, B., 109, 113
Bristol and Its Environs, 40
British Museum's books, 48
Büchner, G., 115

Burford's Panorama in Leicester
    Square, 129

C
Callan, N., 14, 22
Centre d'Etudes Biologiques de Chizé
    (CNRS), 91
chapeaugraphy, 61–2
    'The Big Boot Dance,' 71
    capital exhibitions, 68
    comic pleasure, 71, 77
    'The Cowboy,' 68, 69
    evolutionary theory, 73
    French clown Tabarin, 67
    hat-based gags, 72
    hats, *as props*, 71
    Hercat, 68–70
    theatrical culture, 72
    Trewey, 67–8
    Victorians and Edwardians, 71
    Woodin, England, 67
Chater, R.D., 68–70
cloning, 112
clowning, 62
coil's capabilities, Gassiot, 15–16
Colosseum, 18, 20, 128–9
Cook, J., 40
Cook's Handbook for London, 40–2
    arsenal, assessment of, 46
    gazetteer of London sights, 42
    tourist's scientific requisites, 41
Cook, T., 40
Copenhagen, 107–9, 112
copper wire, 13–14, 22, 24
costumes, 61–2, 72, 118, 129
'The Cowboy,' 68, 69
Crease, R.P., 61
'the Creeps,' 94
Crookes, W., 27, 128
Crystal Palace, 40, 42, 44, 45, 128
cultural performances, 5–6

D
Darwin, C.
  children, experimental
    subjects, 66
  Darwin bicentennial, 2009, 107
  HMS *Beagle*, 39
  ideas of origins, 72
  laughing lunatics, reports, 64
  On the Origin of Species, 108
  scientific research, 39
De Certeau, M., 37, 51
de-hierarchization of theatrical
    means, 114
de la Rue, W., 26
De Marinis, M., 114
Demastes, W., 106
descending incongruity, 63, 74
The Diary of a Nobody, 71
Dickens, C., 2
digital media and technology, 6–7
A Disappearing Number, 110
discharge phenomena, 15, 26
Domani [Tomorrow], 112
Dugas, L., 64

E
electric arc lights, 20
electromagnet, 15, 21–3
electromagnetic induction, 22
emotional labour, 60
emotions, 63–6
evidence-based theories, 3
evolutionary theory, 73
*The Expression of the Emotions in Man
    and Animals*, 64
Ex Vivo/In Vitro, 110
  concept of infinity, 111–12
  kind of objectivity, 112–13
  process of selection, 112
  process over product, 110
  reproductive technology, 110–111

  theatrical project, *Domani
    [Tomorrow]*, 112

F
face chalks, 68
Faraday, M., 13, 22
film-making, natural history
  'Blue-Chip documentaries,' 86–7
  'a literary technology of virtual
    witnessing,' 85
  MOD, 85
  nature films, 85, 87
  property of skill, 85, 98
  strategies of disclosure and
    concealment, 85
  technical skills, disclosure of, 85
  trained performers, 86
  wildlife film-making, culture of, 84, 87
film-making, staging
  bird migration phenomenon, 91
  imprinting theory, 90
  MOD, 90–1
  pelicans flying in 'V' formation, 91
  trained and wild birds, 91, 92
  wildlife film, 92
Finnegan, D., 37
The Flight of the Snow Geese, 86, 92–3, 97
fMRI scanner, 60
Frayn, M., 107–8, 115
Freud, S., 64, 73
Frith, H., 43–4
The Frozen Planet, 84, 97
Fyfe, A., 37

G
gargantuan Monster Coil, 20
Gargery, J., 71
gas battery, 23
gases, electrochemical polarity, 26
Gassiot, J.P., 13–16, 23–8

Gassiot's cascade, 21, 128–9
   induction coil, 13
   knowledge-making process, 14
   spectacle, Victorian visual culture,
      13–14
Geissler, H., 17, 26
Geissler tubes, 17, 26
geological tourism, 41
geology, 37
giggling, 63–4
Gilbert, D., 43
Glasgow theatre, 131
glass-making skills, 24
Goffman, E., 5, 38
Goodall, J.R., 106, 107
Gouyon, J-B., 6, 83–101, 128–9
Great Expectations, 71
Grove, W.R., 15, 23–4, 26–7
guffawing, 63
guidebooks, 37–41, 43–9, 51–2,
   128–30

H
habitus, 66
Harrison, T., 119
hat business
   boy-in-the-hat routine, 76
   child wearing a man's hat, 74
   'comic scene in the nursery,' 75
   descending incongruity, 74
   grotesque or ill-fitting forms, 75–6
   The newer psychology, 75
   sight-gag, 74
   universally comic and funny, 73–4
   Vorstellungsbewegung or mental
      movement, 74
'The Haunted Man,' 2
Henry, J., 23
Hercat, 68–70
Herschel, J., 17, 19, 43

high-intensity electricity, 15
Hilbert's Hotel ('Hotel Infinity'), 111
Hirst, T.A., 126
Hochschild, A.R., 60
Holton, G., 4
Hunt, L., 51

I
illusion, re-staging, 1–2
illusions of reality, 19–20, 130
imagination and science, 2–4, 6–7
imperialism, 44
imprinting technique, 88, 90, 93
induction coil, 13–16, 21–5, 27
infants laugh, 59–60
Infinite Replication Paradox, 111
Infinities, 111–12, 119–20, 128
innovations of science, 17, 50, 119, 128
instrumental practices and techniques,
   21–2
insulation, 24
interdisciplinarity
   creative misprisions, 119
   Infinities and Ex Vivo/In Vitro,
      119–20
   inventiveness, 117–18
   liveness of theatre, 118
   logics of, 119
   public engagement, 116
   real-life scientists and events,
      depictions of, 115
   research funding and evaluation, 117
   scepticism, 117
   science-in-theatre, 115–16
   'seeing what sticks' model, 116
   Snow's 'two cultures,' 116
   theatre and science, 116, 118–19
   transdisciplinarity, 118
interdisciplinary perspective, 130
Interdisciplinary Science Reviews, 4, 106

international exhibition, Edinburgh, 40
Introductory Text Book of Geology, 41

K
Kearton, C., 95
Kember, J., 107
Kitchener, W., 130
Kronecker *vs.*Cantor, 111

L
Latour, B., 2–3, 61, 84–6
laughing, 63–4, 67
laughter, science of. *See* hat business
Lefebvre, H., 37
Lehmann, H-T., 114
Leviathan and the Air Pump, 4
Life of Galileo, 109
Lightman, B., 3, 37, 107
Lipps, T., 74, 75
Little Tich, 71–2
Living Forever, 111, 120
Livingstone, D.N., 36, 39
Lorenz, K., 88, 90
Lotze, R.H., 63

M
Macbeth, 47
magneto-electric device, 22
making-of documentaries (MOD)
'a work of translation,' 86
evidence of film-makers'
capacity, 98
hybrids of nature and culture, 86
knowledge producers, 87
pre-making-of era, 96
staging, 88
material and cultural resources,
21, 24–5, 27

Maxwell, J.C., 126
McDougall, W., 64
mechanical objectivity, 65, 76
mediated science
biography, 108–9
cultural receptivity and
predisposition, 109–10
Darwin-FitzRoy period, 108
natural selection, 108
scientists, biographical dramas, 107–8
Meredith, G., 64, 73
military Hospital, 36
military research site, 46
Milsom, A., 41
Mnemonic, 110
mobility, 3–4, 6, 51, 113
MOD. *See* making-of documentaries
(MOD)
Molière, 65
The Monster Coil's dimensions, 20
Moorish palace, 129
Morning Chronicle, 130
Morse, S., 23
Morus, I.R., 4, 6, 11–31, 37, 60, 85,
98, 107, 128–30
Myers, F.W.H., 62

N
naturalists
The Flight of the Snow Geese, 93–4
flocks of snow geese, 92
imprinted birds, 'the Creeps,' 94
Survival, 92–3
tame/wild distinction, 94, 96
Winged Migration, pre-production
phase, 93
natural selection, 108
nature films, 85, 87
Nead, L., 48, 50
neutral mood, 60

'New Invisible Nose,' 68
Newton's bridge, 51–2
Nowotny, H., 117

O
O'Connor, Ralph, 37
On the Origin of Species, 73, 108
ontology, logic of, 119
Open Fields: Science in Cultural Encounter, 119
organ transplantation, 112

P
Panopticon, 128–9
paper-folding, 69, 70
Park, M., 39
Parnell, P., 107
passions, concept of, 63
Pepper, J.H., 1–2, 19, 130
Performance and Evolution in the Age of Darwin, 106
performance studies, 38
    collisions and collaborations, 7
    competitions and prizes, 7
    cultural performances, 5–6
    digital media and technology, 6–7
    issues, 4
    political commitment, 6
    research institutions, 7
    scholarship, 6
    science television, 7
    visual, analogical, and thematic, 4
performativity of laboratory techniques, 61
period's theatrical culture, 61–2
Perrin, J., 85, 91
personal electrical halo, 17
Peyret, J-F., 109–11, 128
photography, 27
Planet Earth, 84

plurality of causes, 64
polytechnic's hydro-electric machine, 20–1
postdramatic performance, 114–15
prejudices of sense, 19
pre-making-of era, 96
primitive societies, 66–7
props, 62, 68, 71, 113, 129
public scientific self, 12
puppets, 61

R
Reid, T., 18
relief theory, 63
Republic, 60–1
Ribot, T.A., 64
romanticism, 39
Ronconi, L., 109, 111, 128
    acting and design, 113–14
    audience mobility, 113
    Biblioetica, 128
    defamiliarization, 114
    fact of omission, 113
    Infinities, 128
Routledge's *London and its Environs*, 42
    city's global significance, 43
    historic and contemporary scientific sites, 44
    monument-less cityscapes, 44
    social progress, 49
    Woolwich Arsenal, 46
'Royal Arsenal,' 46
The Royal Institution, 42, 44, 45, 125
Royal Observatory, 36
Royal Polytechnic Institution, 129
rubber cultivators, 24
Ruhmkorff coil, 23
    coil's capabilities, Gassiot, 15–16
    discharge phenomena, 15
    high-intensity electricity, 15

induction coil, 14–16
spectacle of cascade, 16
Ruhmkorff, H., 15

S
Saxton, J., 22–3
Schechner, R., 6, 38, 87, 98
science and performance, 3–4,
    98, 106
science and theatre, 4, 20, 105–6,
    114, 118, 127
science humanities, 3, 7
science in Victorian London
    British urban culture, 38
    tourist locations, 43
    travel guidebooks, 43–4, 52
    travel guides, 42–3
science scholarship, 2–3, 6–7, 37,
    52, 106
'science's movement *over space*', 37
science television, 7
scientific facts, creation of, 60–2
scientific spaces, 37–8
scientific spectacle, 20, 24, 28, 130
scientific travel writing, 39
second law of thermodynamics, 108
sensational spectacle, 18–19
sensible impressions, 17
shadowgraphy, 69
Shakespeare, 44, 47, 65, 129
Shepherd-Barr, K.E., 4, 6, 105–22,
    127, 130
The Shows of London, 106
*A Six Month Tour through the North of
    England*, 39
Smith, T.W., 6, 59–80, 106, 128
sniggering, 63–4
Snow, C.P., 60–1, 116
South Kensington museums, 42, 44,
    49
sparks, experiments, 23

spectacle
    artful illusions of reality, creation,
        19–20
    demonstrations, 27–8
    electric arc lights, 20
    experimental, 23
    experimental display, 25
    Gassiot's cascade, 17
    panoramic and dioramic, 129
    philosophy of sensation, 17–18
    technologies of display, 20–1
    use of, 128
    visual deception, 19
    visual sensation, 18–19
Spencer, H., 62–7, 73–4, 76
Square Rounds, 119
Staging Consciousness, 106
stand-up comic, 60
Stokes, G.G., 27
Stoppard, T., 107
Studies of Childhood, 65
Sturgeon, W., 21
Sully, J., 61–7, 73–7

T
Tabarin, 67, 72
telegraphy, 23–4, 28, 42
The Tempest, 44
theatrical and scientific practices,
    60, 61
theatricality and intermediality
    acting and design, 113–14
    Biblioetica, 113
    concept of choice, 113
    Ronconi's theatre work, 114
'Thing Theory,' 72
thinking, centrality of, 36
Thomson, J.J., 27, 128
Time Travel, 111
The Town, 52
transdisciplinarity, 118

travel guidebooks, 52
  advertisers, 41
  agricultural territories, 40
  Bradshaw's London guide, 40
  industrial events, 39–40
  science, 41
  scientific travel writing, 39
travel guides, 38–9
  content of, 37
  science in Victorian London, 41–2
Travels to the Interior Parts of Africa, 39
Trewey, 67–8
Trumpery, 107
Turin Olympics, 112
'two cultures,' 60
'Tyndalising my imagination,' 126
Tyndall, J.
  Alpine glacier, demonstration of, 125
  combinations of apparatus, experiments, 126
  science and culture, 127
  theatre and science, characteristics, 128
  worship of science, 128

U
Unarmed Hunters, 95–6
unmediated science plays
  illegitimate forms, 106
  interdisciplinarity, 107
  legitimate mainstream stage, 106
  performance studies, 107
  plays and performances, scores, 106
  theatre scholarship, 106

V
vacuum tubes, electrical discharges, 23, 26

Varley, C., 26, 128
ventriloquism, 67–8
Victorian physics, 12, 27–9
Villars, P., 43–4, 49
Von Helmholtz, H., 63
The Voyage of the Beagle, 39
A Voyage Towards the South Pole, 39

W
Walden, 111
Watt-Smith, T., 106
Weimerskirch, H., 91
Westminster Bridge, 49–51
wildlife films, 84–8, 96–8, 128
Willis, M., 1–8, 18, 35–56, 60, 128, 130
Winged Migration, 85
  bird migration, natural phenomenon of, 88–9
  film-makers' property of skill, 89
  film-making, scientific enterprise, 90
  imprinted birds, 88–9
  knowledge production, 90
  MOD, 89, 96
  spectacular effect, 88
  trained birds, 89
Withers, C., 36, 39
Wolff, T., 107
Woodin, W.S., 67, 72
Woolwich Arsenal, 42, 46, 128
Wordsworth, 51
Woyzeck, 115
Wyld's Globe, 129

Y
Young, T., 39

Z
Zoological Gardens, 36, 42, 48